Management for Profession

The Springer series *Management for Professionals* comprises high-level business and management books for executives. The authors are experienced business professionals and renowned professors who combine scientific background, best practice, and entrepreneurial vision to provide powerful insights into how to achieve business excellence.

More information about this series at https://link.springer.com/bookseries/10101

Peter Bebersdorf • Arnd Huchzermeier

Variable Takt Principle

Mastering Variance with Limitless Product Individualization

Springer

Peter Bebersdorf
AGCO, Fendt
Marktoberdorf, Germany

Arnd Huchzermeier
WHU – Otto Beisheim School of Management
Vallendar, Germany

ISSN 2192-8096 ISSN 2192-810X (electronic)
Management for Professionals
ISBN 978-3-030-87172-7 ISBN 978-3-030-87170-3 (eBook)
https://doi.org/10.1007/978-3-030-87170-3

This Springer imprint is published by the registered company Springer Nature Switzerland AG.
The registered company address is: Gewerbestrasse 11, 6330 Cham, Switzerland

For Annette, Lotta, Paul and Mats. For your love, strength, warmth and the future with you.
Peter

For Claudia, Alexandra and Stefanie. For your love, patience and support.
Arnd

Foreword

While the west was focused on better control and optimization of assets, Toyota was discovering how to build and manage integrated systems for creating value. Key principles included linking activities in a continuous flow driven by the pull and mix of demand. They also developed a Heijunka methodology for levelling the mix over each production cycle.

But could this be used where the work content for different products varied considerably. Early on the clothing industry did just that with what became known as the "Toyota Sewing System." This book describes the theory and practice in a manufacturing setting. This also creates a common rhythm or Takt time enabling everyone to see progress to plan in real time.

However, Toyota also discovered that integrated systems only work if each step can be performed "right first time—on time—every time." To do this they also engage those carrying out the work in the detailed design of the work steps for each type of product, to maintain quality at source, and to respond quickly to the frequent interruptions that arise throughout any integrated system.

This experience lays the basis for continuous detailed improvements to the system, are valuable in responding to changing circumstances and for teams designing next-generation products and processes. Indeed, they lay the foundation of a common problem-solving language for operatives and engineers throughout the organization. This book describes an important building block in designing responsive and adaptive systems to meet the challenges of the future.

Lean Enterprise Academy
Llandudno, UK

Daniel T. Jones

Contents

1 Introduction

1.1 Our Motivation for This Book

Variable takt—at first, this term sounds like a contradiction in terms for many who have worked in, with, or for assemblies—or still do.

Because the takt is fixed— or not?

A takt is the division of a rhythmic sequence into equal units, so in linguistic understanding. In production, it is much more than that for the employees involved. The takt or takt time is one of *the pivotal points* around which everything revolves in assembly. But does it really have to be fixed, i.e., be constant? Or would it not be better to keep the workload on the employees constant, regardless of which product-mix is assembled with which options. This is not possible in a fixed takt.

With the variable takt and the VarioTakt, we would like to present an assembly concept in this book which has been known in science and theory for more than 50 years (Kilbridge & Wester, 1962; Dar-El, 1978) but has not yet found any significant implementation in industry. Wrongly! We would like to contribute, encourage, and support the much wider use of this assembly concept. We will use one of the pioneers and innovators in the large-scale use of variable takt, the tractor production of Fendt, as an example in many places in this book. With it, we have succeeded for the first time in extracting the essential adjustments for successful implementation and presenting them in an understandable way—copying is encouraged!

There are reasons for the reluctance to implement variable takt times to date, which we will discuss. However, the current and future challenges (Sects. 2.1 and 4.1) of the manufacturing industry, in particular the assembling industry, demand new concepts. In this context, we do not believe in the end of flow assembly, as already propagated by numerous consulting companies with the hope of lucrative contracts. On the contrary, with variable takt and in combination with other assembly concepts, as we present in Chaps. 7 and 8, the numerous advantages of continuous flow assembly (Sect. 2.1) can be expanded even further. In the course of this book, we will be able to show that

P. Bebersdorf, A. Huchzermeier, *Variable Takt Principle*, Management for Professionals, https://doi.org/10.1007/978-3-030-87170-3_1

the assembly of different and varied products in an assembly line in a fixed takt cannot represent a continuous flow of workload! When using variable takt, however, an even flow of workload for all employees in the assembly line is achieved almost automatically

—and without incurring additional effort in sequencing (Sect. 3.2). On the contrary, sequence constraints can be resolved.

It is not the product itself or the production program that has to be standardized in order to level the (assembly) processes! The assembly system itself must be able to handle the variance in different products, to level the load. Variance in and between products can be maintained or further increased for the customers.

With the exception of a few theoretical considerations on variable takt, no documentation, guidance or experience on the introduction and application of variable takt is available. We would like to close this gap by documenting the basics of variable takt in this book.

We ourselves are assembly professionals—one of the authors has been working in planning and managing assemblies since the start of his professional life. We would like to pass on our experience and knowledge and therefore pay attention to what we write between the lines. We would like to help other companies, planners, managers, but also students and scientists to understand and implement variable takt or to motivate further research in this subject area.

We want to help ensure that assembly operations and the jobs they provide can continue to be run economically in our home countries. The manufacturing industry should continue to play a key role in society in the economically highly developed regions of Europe and North America and not have to migrate to low-wage countries. It should be possible to assemble even more individual products for customers economically and under the best working conditions for the employees in assembly.

1.1.1 Assembly in Takt: A Recognition!

When visitors are greeted at an assembly plant, one of the first questions they usually ask is: What's your takt time? Or less technical: How fast do you work here? For the employees concerned on the assembly line, it is the rhythm in which their world turns for 7–8 h a day, 5 days a week. Much research has been done and implemented in manufacturing companies over the past decades to make working in short takt times less physically and mentally stressful for the employees involved. In most assembly lines, primarily in the automotive environment, value is added by employees, i.e., by people. Human labor, and especially work with one of nature's most complex tools, the human hand, will continue to play a crucial role in assembly operations in the years and decades to come. Whereas most manufacturing techniques can be highly automated, it will still take some time before the variability and sensitivity of a human hand and the flexibility of an employee can be replaced.

The assembly and the required assembly skills are often underestimated!

Skills and abilities in the classic fabrication areas are often classified as very complex and high-value, which they are in established industrial companies. However, efficient and effective assemblies that have to prove themselves in a tough competitive environment are in no way inferior to this. Instead, they have a different focus. Because fabrication machines usually appear more impressive and less comprehensible to outsiders, at least at first glance, these areas are classified as more demanding. If visitors to an automotive assembly line observe an experienced worker doing his/her job, it usually appears calm, controlled, not at all hectic and actually quite simple. It is not often that one hears "Well, I could do that too." But it is precisely this impression that requires the highest level of professionalism.

- It is crucial how parts and tools are picked up and held—with the left hand, tilted, slightly slanted, with preload or without. Here small differences can cause big effects, professional training and a lot of experience are crucial. These skills, these subtle crucial differences in handling must be documented, standardized, and passed on.
  One of the most common techniques in assembly, screwing, seems very simple to outsiders, because everyone has used a cordless screwdriver at least once. However, process-capable screw connections from 3 Nm to 450 Nm in short takt times require a high level of professionalism.
- Despite massive improvements in reducing physical stress during assembly activities in recent decades, assembly remains a physical process. This process requires not only physical fitness but also a high quality in the execution of physical movements in order to be able to perform them permanently, in extreme cases for a lifetime.
- The line runs and runs . . . maintaining inner calm while the workpiece continuously moves away from one's own workstation is one of the most decisive skills to be able to work permanently in an assembly line. Ensuring this, even in the event of minor errors or malfunctions, is a fine art.
- But not only the activities in the assembly line have to be carried out professionally, the organization and the daily management are crucial. In assembly lines, activities are mainly carried out by human individuals, with their individual experiences, skills, attitudes, and also mistakes. This must be taken into account in the daily management and design of the assembly. The challenge is to use the individual differences of the employees but to prevent them from having a negative influence on the efficiency and effectiveness of the assembly.

Machines and equipment deteriorate with the time of their operation, they wear out and need to be replaced. The fantastic thing about the employee is that he/she improves him-/herself and his/her activity independently . . . up to a certain limit, with good management even quite a bit beyond. If this process is professionally supported and coached, a superior continuous improvement process is created.

The takt is one of the most important parameters when designing and changing assemblies and their processes. Keeping it constant for as long as possible is considered the holy grail, because changing it usually involves a lot of effort and a loss of quality of the products to be assembled. Not to be underestimated in the case of a change of takt is the stress and the many possibilities for conflict among colleagues, managers, and works council involved—even if these influences are difficult to assess in monetary terms. Thus, the fixed, unchanging takt forms the dilemma from which many of the new assembly concepts try to break away (Sect. 7.2).

Manual assembly and the people employed in it will continue to be an important pillar of our industry and thus of our society in the coming decades. Assemblies must be professionally executed, planned, and managed, variable takt and the VarioTakt can help to meet the changed requirements (Sect. 1.2).

1.1.2 Practice and Science: The Target Groups

With this book, we are undertaking a balancing act. Because we combine business with science, practice with theory. We are certain that good entrepreneurship and professional production management are based on scientific findings and will be even more successful if they make greater use of them. And the other way around! In this context, the question of chicken and egg does not arise. Because entrepreneurial implementations can also initiate or drive scientific work in the exploration of variable takt. This is exactly the way this book was written. In Sect. 4.3.1, we briefly describe the genesis of the VarioTakt at Fendt. As part of the INSEAD-WHU Industrial Excellence Award (IEA, formerly "The Best Factory") in 2017, the first collaboration between WHU - Otto Beisheim School of Management and Fendt's tractor production took place. The joint goal is to research the framework conditions, the mode of operation, and the success factors of variable takt and to make them accessible to a broad public from practice and science.

So, this book has different target groups and we try to do justice to each as appropriately as possible:

- For planners of assembly lines, we would like to describe the mode of action and the implementation steps when introducing variable takt. Often, these target groups are still stuck in existing thought patterns, in which a variable takt does not exist, perhaps even must not exist. Here we would like to convince with many positive examples.
- For students of production science or Operations Management, we would like to introduce them to the operation and nature of an assembly line. Generate an understanding of the real practical challenges in everyday business and explain in detail how variable takt works.
- We would like to motivate scientists to do much more research on methods and concepts to be able to operate a variant-rich assembly efficiently and effectively. For this target group of our book, we hopefully generate more questions than answers. For example, there are still many gaps to be filled in the area of selecting and determining takt time groups. In addition to the use of variable takt in the phase of balancing of an assembly line, which is the main focus of our work, we outline in Sect. 8.6.4 an approach to use the principle of variable takt also in the

control phase of an assembly, the sequence planning—series-sequencing algorithms adapted to this and first validations would be desirable.
- Our actually most important target group: the assembly managers! Because they are the decision-makers in the introduction of variable takt. Without their "go" there is no implementation and without implementation, there is no dissemination and no further development of the concept of variable takt. With many arguments and positive examples, in the focus in Sects. 4.5, 4.8, and 7.1, we try to motivate for implementation.

Maybe the reader should try a little thought experiment while reading: Put yourself in the place of the other target group at one or the other point, which may not be written appropriately for your thinking or understanding. Scientifically oriented readers should also deal with the practical contents and experiences from Sects. 2.1, 3.1, and 4.1; decades of experience from practical applications of Lean Management are hidden here. Practitioners should, despite all theory, try to understand the intention of the generalizing approach in Chap. 5, because this way the practical knowledge can be transferred to other systems or the existing system can be improved. There is the chance of an even deeper understanding of variable takt. This is exactly the way we have enjoyed working on variable takt and VarioTakt in recent years, and an even deeper understanding of each other and of the subject has emerged.

1.1.3 Questions that Guide Us in This Book

Before we started working on the topic, we did what every good manager and scientist should do before starting his/her business. He/she asks his/her customers. Through many lectures and benchmark visits on the VarioTakt and through exchanges with scientific colleagues, we collected the most important and interesting questions that we would like to answer in this book. We additionally add content that is not obvious at first, but which we believe to be highly relevant for mastering variant-rich assembly. This is how, among other things, Chap. 3 "Heijunka—fast as a tortoise" was created.

In summary, our guiding questions are:

Chapter 1:

- What are the global and sustainable developments that require assemblies to master a significantly increased variance?
- How and where does variable takt fit into the design, planning and control phase of an assembly?

Chapter 2:

- What are the advantages of producing in a continuous flow assembly line?
- Why and when do fixed takt concepts reach their limits?

Chapter 3:

- What is the benefit of levelled production?
- What approaches can be used to level the load of a build-to-order assembly?

Chapters 4 and 5:

- How does the variable takt work—like the VarioTakt from Fendt?
- What are the advantages of using the variable takt?
- What are the requirements for using the variable takt?
- What are the experiences with the introduction of the variable takt? What are the success factors?

Chapter 6:

- How can product design help master variance and reduce takt losses?

Chapters 7 and 8:

- With which assembly concepts does the (automotive) industry try to master the increased variance of products?
- How do the variable takt and the VarioTakt compare with competing assembly concepts?
- How can the variable takt be combined with these concepts?

Chapter 9:

- How does the production system of the future look like?

1.2 Global Trends Pose New Challenges for Assemblies

Manufacturing companies and their associated assembly lines have had to and still have to adapt to continuous change. Whereas these were initially strongly regionally driven, today's assemblies are in global competition with globally branched supply chains. Three current global developments are forcing changes in existing assembly concepts.

- Increasing customer individualization and centricity—from mass production to mass customization to mass individualization
- Hedging global risks—corporate resilience
- Disruptive innovations in products—through technology diversification in products

1.2.1 Increasing Customer Individualization and Centricity

Not only in automotive engineering, but in almost all industries, there is an increasing trend toward individualized products, in the extreme toward individual one-off products. In classic craft production, before the industrial age, this was still the rule (Fig. 1.1). After the start of industrial individual production, mass production increased rapidly in the following decades. Certainly, the most famous representative at that time was the Ford company with its T-model, which was available in any color, the main thing was that it was . . . the reader knows. The individualization of products disappeared. Taiichi Ohno and the Toyota company realized in the following decades that American automobile production was clearly superior to them. Their greatest achievement was to realize that they could not and should not copy American principles. The triumph of Lean Manufacturing took off and made Toyota one of the most successful automobile manufacturers of all times. Toyota's approach was able to take advantage of mass production and still produces multiple models or products on one line. First approaches of variable takt production are also reported by Toyota. In our opinion, Toyota concentrated too long on the standardization of its products to avoid variance in assemblies, instead of mastering customer-oriented variance. In the meantime, not only the automotive sector, but most of the manufacturing industry has realized that their future lies in the production of individualized and personalized products. Mass production became mass customization (Pine et al., 1993; Kotha, 1995; Da Silveira et al., 2001) and permanently changed production systems in the last decades (Fig. 1.1). Customers want products tailored to their individual needs, and as a result, product and option variants have increased many times over, significantly exceeding pure volume increases in all industries (Wildemann, 2020). No niche is too small not to be occupied by an individualized product. In addition, this individualization opens up the possibility

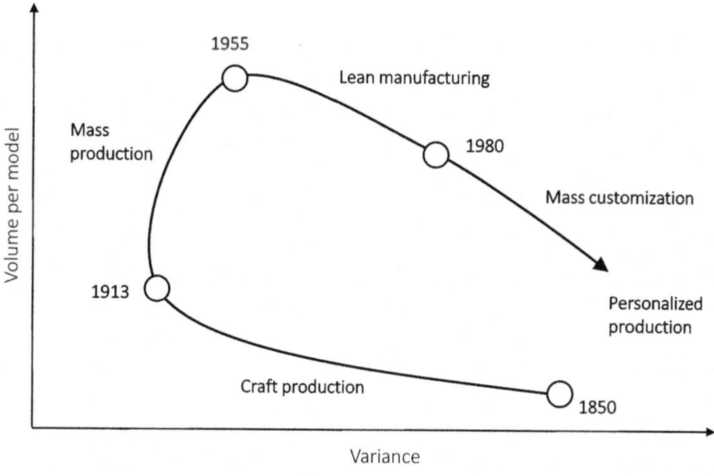

Fig. 1.1 Variance transformation in production (adapted from Koren, 2010; Koren 2021)

Table 1.1 Comparison—mass production/customization/individualization (adapted from Koren, 2021)

	Product architecture	Product type	Role of the customer
Mass production	Uniform	Identical products	Select product
Mass customization	Modular	Products with options	Configure product
Mass individualization	Open structure	Products designed by the customer	Design/co-create product

for companies to further differentiate themselves from competitors with high margins. In Sect. 2.4, we will discuss the effects of these developments on assembly systems in more detail. Assembly, usually the last part of the manufacturing chain, has the greatest task in the individualization of products.

If the ideas of Fig. 1.1 continue, the flexible production of a single individual product in a local factory will become the task to be mastered by many customer-centric companies in the future. In Table 1.1 we show the evolution from mass production to mass customization to mass individualization (Koren, 2021).

The ubiquitous expansion of the Internet into almost all areas of life and the strong expansion of a wide variety of social networks make it easy for customers to generate and exchange new unique product ideas and place orders with companies (Jiang et al., 2016). The loop in Fig. 1.1 is closing, but from the market of individual products for a wealthy few to the market of individual products for everyone (Fader, 2021; Koren, 2021).

In his forecasts, Koren (2021) sees two different production structures for this development:

1. Factories that produce individual single products using highly flexible and automated manufacturing and assembly technologies. Automated matrix structures (see Sect. 7.2) or additive manufacturing technologies can be such solutions.
2. Assemblies that produce individualized modules (e.g., cockpits of passenger cars). Open design structures that are assembled on standardized basic modules (Koren et al., 2013).

Individual products that the customer designs him-/herself cannot be manufactured in a globally concentrated manner far away from the customer. Interaction with the customer is essential, and customers also expect delivery of their individually designed product in the shortest possible time (Koren, 2010). This is not possible with production on the other side of the globe; globally distributed assemblies with large product portfolios are the result.

Product individualization or personalization will become a decisive competitive advantage not only in industrial companies. Ben-Jebara and Modi (2020) showed that in the future up to 50% of products in the pharmaceutical industry can be individualized and thus significantly improve the financial performance of companies.

Increasing individualization means that assembly lines must be able to handle a greater number of different products with a wide range of options. As the variance of products increases more than the total volume, the number of units per product model decreases. Single-product assembly lines can no longer be operated economically.

This development is supported by an increasingly strong desire among customers to use our resources sustainably and responsibly (Whelan & Kronthal-Sacco, 2019) and the resulting desire to produce products locally (Lund et al., 2020). Volumes are no longer bundled globally.

Greater individualization of products leads to stronger, seasonal fluctuations in the products and option variants in demand. In order to level out this fluctuating demand, it is an advantage to be able to manufacture the different products in one assembly system. The increasing customer focus on more and more individualized products is forcing companies to produce more and more different products with even more options in one assembly system. Assembly in a fixed takt reaches its limits here at the latest (see Chap. 2).

1.2.2 Hedging Global Risks

Through a globally networked production structure, regional advantages such as lower wage or energy costs, technological clusters or proximity to raw materials can be used specifically to achieve economies of scale through concentration. In a world that is becoming increasingly uncertain and unpredictable, this networking, usually with minimal buffers, now also represents a risk for these companies (Winton, 2021).

Individual events, such as the blockade of the Suez Canal in April 2021 by the freighter MV Ever Given, show how vulnerable supply chains are at some of the chokepoints (literally) on the world map. But potential military conflicts in regions that are critical to world production, such as China and Taiwan, also raise the pulse of many supply chain managers.

Increasing nationalism, as well as the shift in global economic importance from North America toward Asia, lead to an increase in trade barriers such as tariffs or import restrictions. The local parameters under which production networks have been set up change quickly and more unpredictably due to political measures. Labor cost advantages of individual countries can quickly disappear, currency risks amplify these effects.

An increase in natural disasters, triggered by climatic change, can bring entire regions to a standstill in the short term (Cohen et al., 2018). The current Corona pandemic, with strong local differences, shows the globally interconnected world its limits (Altman, 2020). Resilience (Cohen et al., 2021) has been one of the most discussed buzzwords or concepts in current management discussions at least since the Corona pandemic.

The failure of individual local points on the global map brings the entire network to a halt (Winton, 2021; Gitlin, 2021). A trend toward localization of supply chains emerges (Lund et al., 2020; The Economist Intelligence Unit, 2020), i.e., a trend toward the distribution of final assemblies on the global map.

The increasing ecological awareness of many customers and now also investment companies and pension funds support this global distribution of production.

The combination of all these factors forces companies to make their production network more resilient and flexible (Whelan & Kronthal-Sacco, 2019; Ivanov & Dolgui, 2020; Linton & Vakil, 2020). One way to make the production network more resilient is certainly to distribute production more globally (Cherney, 2020); large supply chain and strategy consulting firms are currently focusing heavily on this trend (Busellato et al., 2021). This localization of production initially entails significant cost disadvantages, which must be compensated for (Cherney, 2020).

In order to nevertheless realize economies of scale, production, and in our case assemblies, will be bundled regionally—but no longer globally on a product-specific basis. Smaller standardized assemblies that manufacture different products will be the result (Shih, 2020). These assemblies are thus faced with the challenge of having to map a large number of products with different basic structures, a wide variety of options, and fluctuating assembly efforts. In the course of this work, we will show that assembly systems with a fixed takt are thus pushed to their limits.

1.2.3 Disruptive Innovations: Technology Diversification

The individualization of products is supported by numerous disruptive innovations that lead to technology diversification in products. Electric mobility, for example, is on the verge of a breakthrough. Automobile manufacturers are confronted with offering different drive concepts (Küpper et al., 2018). Due to the associated lower units per variant and the high investments in an automotive assembly, these vehicles cannot be manufactured in their own assembly specialized in their product structure. These technologically different products have to be assembled on one assembly line, large local and global takt time spreads (Sect. 2.6.11) are the result. The temporal and spatial differences in the load on the assembly must be mastered. But it is not only new drive concepts that lead to these extreme takt time spreads; (partially) autonomous driving also requires the assembly of a large number of different control units and sensors, whereas products without this capability do not need these components. This development can be observed not only in automotive engineering; Canyon Bicycles, for example, assembles its product range of racing bikes, mountain bikes, and electric bikes, which has been greatly expanded in recent years, on just one assembly line. In Sect. 5.5, we present this challenge in a case study.

Product innovations are the engine of our economy and the basis of the entrepreneurial success of every company; production and, in particular, assembly must master their effects. Large takt time spreads must be mastered, in fixed takts this is only possible with numerous compromises.

1.2.4 A Conclusion for Assemblies

Giga Factory, under this term Elon Musk and Tesla open their new factories. However, it can be assumed that with Giga Factory, Tesla is aiming less at economies of scale through volume and more at bundling the entire value chain at a local site—from the battery to the complete electric car . . . and perhaps its own electronics as well. Tesla's supply chain is thus likely to be more resilient than that of many established car manufacturers. In addition, the factories for all models, now three, are distributed globally. Perhaps the global trends highlighted are thus leading us in precisely another direction: "localized microfactories" (Hanspal, 2019)—small, globally distributed production units. Already one of the main topics at the World Economic Forum in 2019 (Hanspal, 2019). Some examples of these concepts can be found even in the automotive sector, which is otherwise more oriented toward economies of scale, and that both with small startups such as the British Arrival (Stephenson, 2021) or e.Go Life from Germany. But also the large car manufacturers are flexibilizing their assembly lines in order to be able to assemble more models on a single line, like GM's Factory ZERO or Mercedes-Benz Factory 56 (Sect. 8.5).

At the same time, products individualized according to customers must not be standardized; this competitive advantage must be maintained. Customer variance in assembly must be mastered, not suppressed or planned away.

These developments lead to the increasing challenge to operate assemblies economically without using the classic economies of scale. Through our work, we would like to show ways to maintain and increase economic efficiency despite an expansion of variance in assemblies. Using variable takt and our approach of Mixed-Model Assembly Design (MMAD—Sect. 1.4), the following three main cost types of an assembly with high variance can be reduced:

1. *Costs in assembly operations*—by reducing utilization losses (Sect. 2.8)
2. *Modification and maintenance costs*—for adjustments to volume, product-mix or product design
3. *Investment costs*—due to a reduction of the necessary investments during initialization or changes in the assembly

And what about the quality of the products? "Without quality, nothing is gained"—as experienced producers, we know that the most efficient assembly is also the one with the highest quality of its final products. There is no "or" between quality and efficiency! At many points in this paper, we will point out that the methods and concepts described also positively influence the quality of the end products. For example, the variable takt, despite large variations in the work content of the products, provides a levelled and constant workload for the workers on the assembly line. An even workload is one of the most important prerequisites for achieving it.

4. *High quality* in an assembly line

1.3 Drivers of Variable Takt at Fendt

No change without necessity! One of the essential drivers for change and innovation is necessity (Ohno, 2013). This insight sounds trivial at first, but it is fundamentally anchored in organizations and systems. Without crisis, without problems—no change. Even the Toyota Production System (TPS) has its origins in this, as we explain in Sect. 3.1. The VarioTakt, or its first large-scale implementation, also arose from a necessity. The market for agricultural technology and, in particular, the tractor production of Fendt set very specific framework conditions for an assembly, which can ultimately best be answered with the application of the variable takt. In the following and in many other sections of our book, we will demonstrate this.

1.3.1 Agriculture and the Market for Agricultural Technology

In 2020, a new record was set in the harvest of the main arable crops. Very good yields in almost all regions of the world were one reason for this. The steady expansion and ever-increasing professionalization, and thus increased efficiency, of agriculture are further reasons. The increase in food production is essential for the growing world population, which currently has to be fed around eight billion people. By 2050, this figure will have risen to 10 billion—an increase of almost 25%. And more than 800 million people are still going hungry, according to the Food and Agriculture Organization of the United Nations—a far too large figure, albeit with a declining trend. Agricultural technology, one of the most important suppliers to agriculture, has a crucial role to play here. It must supply agriculture with machines, technology and increasingly also software in order to increase yields, both in terms of quantity and quality, while using increasingly reduced resources. Unnoticed by many, agricultural technology is the technology leader in many areas. For example, autonomous driving in the field with an accuracy of 2 cm has been standard for many years. "Most innovations in recent years can be subsumed under the heading of precision agriculture. One of the most important elements here is the collection and documentation of data, be it the registration of crop condition or fuel consumption per hectare. The end result is an optimally networked and automated process chain with minimal use of inputs and maximum transparency and efficiency for completing operations" (Wiesendorfer, 2016). Above all, the trend toward dispensing with pesticides and fertilizers reinforces the need for precise and data-based agriculture. Digital twins of agricultural areas and farms are emerging.

According to the VDMA (German Machinery and Plant Manufacturing Association), the global market volume for agricultural machinery amounted to approximately 111 billion euros in 2019, with Europe, the Americas and Asia each accounting for one-third. For the year 2021, 120 billion euros are already expected. The five largest agricultural technology groups, John Deere, Case New Holland, Kubota, AGCO and Claas, account for around 50% of the total global market. Germany is one of the largest locations for agricultural technology production and innovation in Europe (Wiesendorfer, 2016). Tractors are the core segment of the agricultural machinery market, representing one-third of the global market volume (Wiesendorfer, 2016). Due to a steady structural change toward ever-larger farms in all regions of the world, the motorization of tractors is constantly increasing. Classic wheeled tractors with over 500 hp are now available on the market.

1.3.2 Characteristics of the Agricultural Machinery Market Tractors as Drivers of the Variable Takt

As an investment market, agricultural machinery, and in particular the market for tractors, is subject to three main trends:

- The dependence on the global economy and the development of producer prices on the global stock exchanges. This results in increased volatility in demand.
- Agriculture, and in particular the distribution of its products, is subject to strong political influences and trends. Trade barriers such as tariffs or export and import bans also have a highly volatile effect on demand for agricultural technology.
- The willingness to invest in agriculture depends very much on your results. Harvests can vary greatly regionally, but also worldwide, due to different environmental influences such as droughts, cold spells, etc.

For example, the market rose by +20% in 2008, fell by −20% in 2009, then stagnated in 2010 and rose sharply by 28% in 2011. The players in the agricultural machinery market have adapted to these sharp fluctuations. Agricultural machinery manufacturers must be able to variably adjust their production, like that of tractors.

In addition, there is a pronounced seasonal effect on demand for agricultural machinery during the year (Wiesendorfer, 2016). Demand for different types of tractors fluctuates greatly over the course of a year. Orders for large tractors rise sharply from November (in Europe and North America) and remain at a high level until the end of the first quarter of the following year. Small tractors for fruit growing or viticulture, or tractors for the dairy industry show other seasonal effects. Production must primarily adapt to the fluctuation in the mix of different tractor models in demand.

Since the total volume of different tractors is significantly lower than in the automotive market, all manufacturers of tractors are forced to produce different models on one assembly line. This is exacerbated by the fact that the different tractor models differ more from each other technologically than is the case with different passenger car models. Since the requirements for the different tractor models differ greatly, e.g., between a tractor for field use and a vineyard tractor, they differ considerably in design and consequently in assembly requirements. An extreme example of the differentiation of tractor models is the Fendt brand, which with its large product portfolio offers an ideal application of variable takt on an assembly line.

1.3.3 AGCO and the Fendt Tractor Brand: Pioneers for the Variable Takt

The AGCO Group is one of the largest agricultural equipment companies with 21,000 employees and sales of 9 billion USD in 2020. In its short history, AGCO has established itself as a full-line supplier in the agricultural equipment market with its five main brands Fendt, Challenger, Massey Ferguson, Valtra, and GSI. With its current "Farmer First" strategy, AGCO is pushing forward on the path to becoming a customer-centric company (mass customization). One of AGCO's key assets: the Fendt brand with its highly customized product range (Fig. 1.2). One of the most successful steps in AGCO's development was the acquisition of the company Xaver Fendt GmbH from the descendants of the former founder in 1997. Tractor production of the Fendt brand started as early as 1930 at the Marktoberdorf site in the Allgäu region, Bavaria, Germany. Since that time, Fendt has developed into the technology world market leader for tractors. The development and use of the VARIO transmission, a hydrostatic-mechanical power-split continuously variable transmission, in all tractor models was one of the greatest innovations in agricultural technology in recent decades. The VARIO transmission, with its mode of operation adapted to the load, is also the eponym of the VarioTakt—but more on that later in the course of the book.

Fendt produced nearly 19,000 tractors in both 2019 and 2020 (Fendt, 2021), despite several weeks of plant closures in 2020 due to the Corona pandemic. As in previous years, Fendt ranked first in the German and European Dealer Satisfaction Index in 2020 (Fendt, 2021) and is the market leader in Germany (agrarheute, 2021), one of the largest agricultural equipment markets in the world. Fendt assembles all tractors, of the eight different series, at its Marktoberdorf site on an assembly line, in total-mix, in variable takt, with the VarioTakt (Fig. 1.3). Without the use of the variable takt, it would no longer be possible for Fendt to economically assemble all

Fig. 1.2 Variance extreme—Fendt tractor models in use

Fig. 1.3 Variance extreme—Fendt tractor assembly

tractor models, from the small vineyard tractor to the 500 HP arable tractor (see illustrations in Sect. 4.1), in one assembly line. A detailed description of the necessity of using variable takt in Fendt's tractor assembly is given in Sect. 4.1.1. Not only with variable takt, but with its entire Fendt assembly system, which we describe in Sect. 7.1, Fendt represents a benchmark for mastering variance in assemblies.

The question is often asked: If variable takt brings so many advantages, why hasn't it been used before? The simplest answer to this question is: It was not necessary! In classic automotive manufacturing, single-product lines have been predominant in recent decades. Variable takt cannot develop its benefits in these applications. In tractor production, and especially at Fendt, the multi-product line has been necessary for decades. Products, in this case tractors, with different basic structure, precisely adapted to customer needs, in small quantities, with large fluctuations in product-mix must be mastered during assembly.

The planners and managers involved were not "smarter" than the rest of the production community, there was simply a need for an extended solution. In the meantime, not only Fendt, but numerous other companies with their assembly lines are faced with the challenge of having to map the most diverse products in an assembly line.

The strength of the variable takt is that it can economically map a wide variety of products with a large spread in assembly effort on an assembly line. Together with the VarioTakt and other elements, as described in the following chapters, variance can be mastered in the assembly of a wide range of industries. Due to the special requirements of tractor production at Fendt, one of the first comprehensive applications of variable takt was developed here.

Due to the increasing individualization of customer demand and an increased global distribution of assembly and extended technology diversification in the products, the application field of variable takt will broaden significantly across many industries. Variable takt will become necessary, if it has not already!

In this book, we document important guidelines that allow the introduction of the variable takt into practice. In this way, we are entering new scientific territory. Our own publications, which have been published in well-established scientific journals over the past 3 years, show the high acceptance of this innovative approach by experienced academic reviewers.

1.4 Mixed-Model Assembly Design: Mastering Variance in Assemblies

Already after our first pages, from Sects. 1.2 and 1.3, the insight follows that in the future, assemblies will have to deal with a greater variance *of* and *in* products. In the future, this variance will no longer be limited to different options that could be outsourced to pre-assembly or external suppliers. If different products are mapped in an assembly line, the variance sits in the basic structure of these products—it is global (Sect. 2.4.2). It is no longer an addition or combination of different option variants. The differences in the basic structure of the products cannot be "outsourced," they must be mapped in the flow assembly, they must be mastered. Mixed-model assemblies are created (Sect. 2.2). A precise differentiation between product and options variance is made in Sect. 2.4.

Assembly is a production system that must be planned (Boysen, 2005), in the long term via the conception of assembly or the assembly line (Sects. 2.3–2.5), in the medium term via line balancing (Sect. 2.6), and in the short term via series-sequence planning (Chap. 3). The determining goal of this planning is to minimize manufacturing costs, whereby assembly, due to its large share of manual activities, is usually a driver of operational costs. Due to the variance between and within the products, takt losses (Sect. 2.8.1) and model-mix losses (Sect. 2.8.2) occur in the operation of the assembly line.

Increasing variance increases the complexity of these planning problems and severely limits their solution space (grey area in Fig. 1.4).

If a product is already fully defined, i.e., developed, this further limits the degrees of freedom, and thus the solution space, for reducing manufacturing costs. Although the approaches of DFA, i.e., Design-for-Assembly (Boothroyd, 2005; Chap. 6), can specifically reduce manufacturing costs, DFA does not focus on reducing takt or model-mix losses (Swist, 2014; Sect. 2.8). The goal of our work is to minimize these very losses—**we want to master the variance in mixed-model assembly.**

A Mixed-Model Assembly Design (MMAD) is created (Fig 1.4).

Figure 1.4 shows a schematic course of losses in the operation of an assembly which can be influenced by the use of further production factors in the operation of an assembly. In order to achieve the highest possible economic efficiency of the production system used, the conflict between these parameters must be reduced (Swist, 2014).

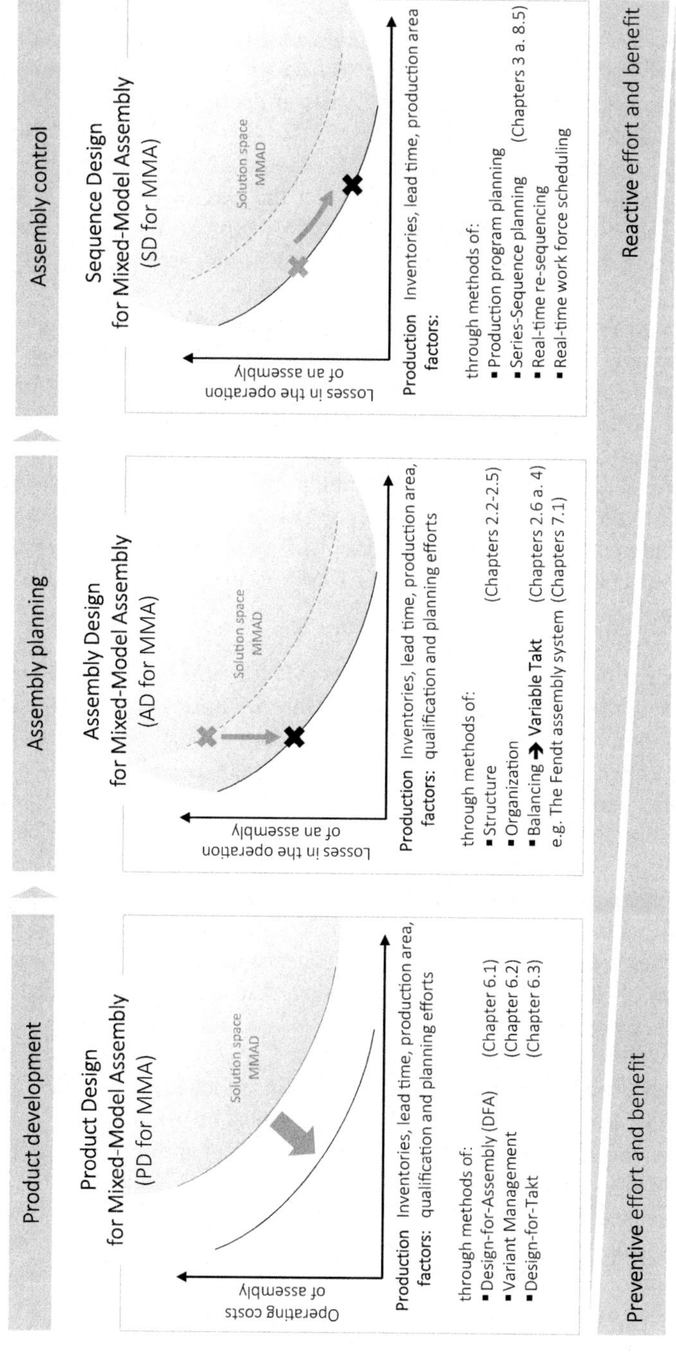

Fig. 1.4 Mixed-Model Assembly Design (MMAD)—adapted from Swist (2014) and Kesselring (2021)

Product Design for Mixed-Model Assembly (PD for MMA) DFA (Boothroyd, 2005) and Variant Management (Wiendahl et al., 2004; Günther & Tempelmeier, 2012) methods can be used to preventively reduce manufacturing costs over a wide range. DFA and Variant Management have so far focused on avoiding variance or reducing overall assembly effort. Reducing costs in the operation of mixed-model assembly, i.e., takt and model-mix losses (Sect. 2.8), is not their focus. This is where Design-for-Takt (Swist, 2014; Kesselring, 2021) should come in, which we develop and describe in Chap. 6. Figure 1.4 depicts that the task of PD for MMA is to increase the solution space for the lowest possible cost in assembly operations without expanding the use of other production factors (grey area in Fig. 1.4).

Assembly Design for Mixed-Model Assembly (AD for MMA) The focus of our book is on assembly design. By means of structures, the organization and above all methods of the takt losses in the operation of the mixed-model assemblies are to be reduced. This is exactly where variable takt comes in. The "X" in Fig. 1.4 symbolizes the current setting point of an assembly. By methods of AD for MMA, in particular variable takt, losses in the operation of the assembly can be reduced without having to accept additional compromises in other production factors. These improvements are possible up to the limit line, which can be moved only by the PD for MMA. A special design of the AD for MMA is the Fendt assembly system, which we describe in detail in Sect. 7.1.

Sequence Design for Mixed-Model Assembly (SD for MMA) Series-sequencing methods can be used to form a product sequence that minimizes model-mix losses in assembly operations, using inventories, lead times, and floor space. This is rather a reactive approach. In Fig. 1.4 the solution point moves down the limit line (reducing losses) but also moves a bit to the right (using other factors). For example, re-sequencing buffers in automotive improve model-mix losses, but true build-to-order sequencing is not possible this way. Workload-oriented series-sequencing (Sect. 3.2) is increasingly finding its way into mixed-model assemblies; if this is combined with variable takt, as we outline in Sect. 8.6.4, additional reductions in model-mix losses are possible. Not only the sequence of products on the assembly line can be controlled, but also the sequence and workspace can be controlled in real time to minimize losses.

In Which Area Should the Focus Be Placed? Since the resources of a company are limited in the vast majority of cases, a decision must be made in which areas of the MMAD the limited resources are used. On the one hand, the preventive approach of the PD for MMA certainly leads to the greatest benefit in the long term, but on the other hand represents a large expense. Research to accurately quantify this cost-benefit relationship would be necessary. However, it can be assumed that for assemblies with a low production volume the approaches of PD for MMA are less suitable. In our estimation, PD for MMA methods have still found the least wide-spread use. Otherwise, it is surprising that the pure reactive approaches of SD for MMA do not dominate either. Our hypothesis: This requires an exact database, which is not available in many, mostly smaller companies. Thus, the focus of work is mostly AD for MMA and thus the assembly planning. The focus is on the selection and design of the appropriate assembly structure, organization and line balancing.

1.4.1 Nature and Structure of This Book

Our book is designed so that each chapter is individually readable and understandable. For this reason, it may happen that we repeat ourselves in some places, we try to minimize this. We often use the Fendt tractor production for our practical examples. We have explained the main reason for this in the previous sections. Very often we continue to use the German automotive industry as an example. The products and basic processes in an automotive assembly are generally well known and understandable. In addition, automotive manufacturing is the innovation driver in the field of assembly technology and assembly concepts—except in the variable takt. It is important to us that all the contents of this book can be applied to any mixed-model assembly, no matter what the industry. We can demonstrate this in an impressive case study of the bicycle assembly of the company Canyon Bicycles (Sect. 5.5).

A picture is worth a thousand words. We deliberately use a very large number of different illustrations to explain our topics. We have made the experience that the functionality and the effects of line balancing of an assembly line can be explained much more effectively and efficiently with pictures. In addition, the reader will remember the pictures longer than the written text.

We are avid followers of Lean Management based on the Toyota Production System and the great thought leaders Ohno (2013), Womack et al. (1990), Liker (2021), and Rother (2009). Their thinking and understanding of managing a production will guide us in many parts of this work.

1.4.2 Structure of the Chapters

In the introduction, we presented the basis for the necessity of switching from fixed to variable takt by outlining current trends and reactions of the manufacturing industry to the changing framework conditions. We briefly explained the specifics of the agricultural technology market and the tractor production of the Fendt brand. Under these conditions, or perhaps because of them, the implementation of variable takt in tractor production developed. Thus, Fendt's tractor production became the global benchmark in the implementation of variable takt and the mastery of variance.

Chapter 2 presents the basis, i.e., the fundamentals to our understanding of balancing in an assembly. We explain the managerial benefits of paced and continuous flow production, subdivide the mixed-model assembly into three types and explain the WATT balancing method. We emphasize the necessary subdivision or distinction between the terms product and options, and the separation of utilization losses into takt and model-mix losses.

In the third chapter, we describe the need for levelled production—Heijunka. We highlight the importance of the three M—Muda, Muri, Mura—from the Toyota Production System (TPS) and compare the TPS approach of Heijunka with the classic methods of series-sequencing. At the end of the chapter, we use real data to compare the levelling of two assembly lines—one in fixed takt and one in variable takt.

The core of our work is Chap. 4. We describe the necessity of variable takt and how it works in detail. We place particular emphasis on presenting the various advantages and experiences with variable takt.

In Chap. 5, we develop different methods to divide the product spectrum of an assembly line into different takt time groups. Qualitative and quantitative approaches are developed and combined. At the end of the chapter, we show with the help of an impressive case study at Canyon Bicycles that variable takt can also unfold its immense potential in a mixed-model assembly outside of automotive assembly.

We describe our Product Design for Mixed-Model Assembly (PD for MMA) approach in Chap. 6. We briefly review the Design-for-Assembly (DFA) and Variant Management approaches and differentiate them from the new Design-for-Takt approach. We elaborate and describe this approach in detail because it most supports the focus of our book, i.e., mastering variance.

Chapters 7 and 8 present our view of the bigger picture. On the one hand, we present the Fendt assembly system and show how successfully Fendt combines variable takt, pre-assemblies and matrix elements. We then take a look at matrix assembly and compare it with variable takt. We show that the break even point of matrix assemblies can be pushed back quite a bit by using variable takt. In Chap. 8, we present the BMW approach to hybrid assembly, a combination of assembly line and matrix, which we believe is very promising and combine it with VarioTakt. The Cycle Module Assembly of the Taycan production is Porsche's answer to master variance, and we also combine it virtually with variable takt. We describe a technologically completely different but conceptually similar approach with Honda's ARC Assembly. Finally, we take a brief look at Mercedes-Benz Factory 56 in Sindelfingen, where workload-oriented series-sequencing is being used in assembly for the first time. We end the penultimate chapter with an experiment. We use a software for automated line balancing in order to compare the effects of fixed and variable takt with changing model-mix.

In the final Chap. 9, we summarize our key findings, encourage implementation, and provide an outlook on potential further developments.

References

Agrarheute. (2021). Tractor registrations: Fendt is market leader in 2020. *agrarheute*. Accessed June 18, 2021 from https://www.agrarheute.com/technik/traktoren/traktorzulassungen-fendt-marktfuehrer-2020-577208

Altman, S. A. (2020). Will Covid-19 have a lasting impact on globalization? *Harvard Business Review Online*, September 2. Accessed April 27, 2021, from https://hbr.org/2020/05/will-covid-19-have-a-lasting-impact-on-globalization

Ben-Jebara, M., & Modi, S. B. (2020). Product personalization and firm performance: An empirical analysis of the pharmaceutical industry. *Journal of Operational Management, 67*, 82–104. https://doi.org/10.1002/joom.1109

Boothroyd, G. (2005). *Assembly automation and product design* (536 p.). Taylor & Francis. https://doi.org/10.1201/9781420027358

Boysen, N. (2005). *Variantenfließfertigung* (294 p.). Springer. https://www.springer.com/gp/book/9783835000582

Busellato, N., Drentin, R., Nair S., Schlichter, M., Kishore, S., & Singh, A. (2021). Rethinking operations in the next normal. *McKinsey & Company Online*. Accessed April 15, 2021, from https://www.mckinsey.com/business-functions/operations/our-insights/rethinking-operations-in-the-next-normal

Cherney, M. (2020). Firms want to adjust supply chains post pandemic, but changes take time; Covid-19 has exposed the risk of farflung production; alternatives like making things at home could raise costs. *The Wall Street Journal Online*, Dez 20, 2020. Accessed December 27, 2020, from https://www.wsj.com/articles/firms-want-to-adjust-supply-chains-post-pandemic-but-changes-take-time-11609081200

Cohen, M. A., Cui, S., Ernst, R., Huchzermeier, A., Kouvelis, P., Lee, H. L., & Tsay, A. A. (2018). OM forum – benchmarking global production sourcing decisions: Where and why firms offshore and reshore. *Manufacturing & Service Operations Management, 20*(3), 389–600. https://doi.org/10.1287/msom.2017.0666

Cohen, M. A., Cui, S., Doetsch, S., Ernst, R., Huchzermeier, A., Kouvelis, P., Lee, H. L., Matsuo, H., & Tsay, A. (2021). Putting supply chain resilience theory into practice. *Management and Business Review*, 16 p. https://doi.org/10.2139/ssrn.3742616.

Da Silveira, G., Borenstein, D., Fogliatto, S., & F. (2001). Mass customization: Literature review and research directions. *International Journal of Production Economics, 72*, 13. https://doi.org/10.1016/S0925-5273(00)00079-7

Dar-El, E. M. (1978). Mixed-model assembly line sequencing problems. *Omega, 6*(4), 313–323. https://doi.org/10.1016/0305-0483(78)90004-X

Fader, P. (2021). *Customer centricity* (150 p.). Wharton School Press.

Fendt (2021). *Facts and figures*. Accessed June 18, 2021, from https://www.fendt.com/de/unternehmen/fakten-zahlen

Gitlin, J. M. (2021). A silicon chip shortage is causing automakers to idle their factories. *Arstechnica*. Accessed February 14, 2021, from https://arstechnica.com/cars/2021/02/a-silicon-chip-shortage-is-causing-automakers-to-idle-their-factories/

Günther, O., & Tempelmeier, H. (2012). *Produktion und Logistik* (388 p.). Springer.

Hanspal, A. (2019). Localized microfactories – The new face of globalized manufacturing. *World Economic Forum Online*. Accessed April 15, 2021, from https://www.weforum.org/agenda/2019/06/localized-micro-factories-entrepreneurs-and-consumers/

Ivanov, D., & Dolgui, A. (2020). Viability of intertwined supply networks: Extending the supply chain resilience angles towards survivability. A position paper motivated by COVID-19 outbreak. *International Journal of Production Research, 58*(10), 2904–2915. https://doi.org/10.1080/00207543.2020.1750727

Jiang, P., Lang, J., Ding, K., Gu, P., & Koren, Y. (2016). Social manufacturing as a sustainable paradigm for mass-individualization. *Proceedings of the Institution of Mechanical Engineers, Part B: Journal of Engineering Manufacture, 230*(10), 1961–1968. https://doi.org/10.1177/0954405416666903

Kesselring, M. (2021). Product Design for Mixed-Model Assembly Lines. Master Thesis, WHU, Otto Beisheim School of Management.

Kilbridge, M., & Wester, L. (1962). A review of analytical system of line balancing. *Operations Research, 10*(5), 591–742. https://doi.org/10.1287/opre.10.5.626

Koren, Y. (2010). *The global manufacturing revolution: Product-process-business integration and reconfigurable systems* (399 p.). Wiley.

Koren, Y. (2021). The local factory of the future for producing individualized products. *The BRIDGE, 51*(1), 100 p. https://www.nae.edu/251191/The-Local-Factory-of-the-Future-for-Producing-Individualized-Products

Koren, Y., Hu, J., Gu, P., & Shpitalni, M. (2013). Open-architecture products. *CIRP Annals, 62*(2), 719–729. https://doi.org/10.1016/j.cirp.2013.06.001

Kotha, S. (1995). Mass customization: Implementing the emerging paradigm for competitive advantage. *Strategic Management Journal, 16*, 21–42. https://doi.org/10.1002/smj.4250160916

Küpper, D., Sieben, C., Kuhlmann, K., & Ahmad, J. (2018). Will flexible-cell manufacturing revolutionize Carmaking? *The Boston Consulting Group Online*, October 18. Accessed April 28, 2021, from https://www.bcg.com/de-de/publications/2018/flexible-cell-manufacturing-revolutionize-carmaking

Liker, J. K. (2021). *The Toyota way* (449 p., 2nd ed.). McGraw Hill.

Linton, T., & Vakil, B. (2020). Coronavirus is proving we need more resilient supply chains. *Harvard Business Review Online,* March 5. Accessed April 28, from 2021 https://hbr.org/2020/03/coronavirus-is-proving-that-we-need-more-resilient-supply-chains

Lund, S., Manyika, J., Woetzel, J., Barriball, E., Krishnan, M., Alicke, K., Birshan, M., George, K., Smit, S., Swan, D., & Hutzler, K. (2020). *Risk, resilience, and rebalancing in global value chains*. McKinsey & Company Online. Accessed January 04, 2021, from https://www.mckinsey.com/business-functions/operations/our-insights/risk-resilience-and-rebalancing-in-global-value-chains

Ohno, T. (2013). *The Toyota production system* (176 p.). Campus Verlag.

Pine, B. J., Bart, V., & Boynton, A. C. (1993). Making mass customization work. *Harvard Business Review Online*. Accessed April 29, 2021, from https://hbr.org/1993/09/making-mass-customization-work

Rother, M. (2009). *Toyota Kata* (400 p.). McGraw-Hill.

Shih, W. (2020). Is it time to rethink globalized supply chains? *MIT Sloan Management Review Online*. Accessed February 14, 2021, from https://sloanreview.mit.edu/article/is-it-time-to-rethink-globalized-supply-chains/

Stephenson, W. D. (2021). Microfactories, not gigafactories, will build it Back better. *Industry Week Online*. Accessed April 15, 2021, from www.industryweek.com/technology-and-iiot/emerging-technologies/article/21161158/microfactories-not-gigafactories-will-build-it-back-better

Swist, M. (2014). *Taktverlustprävention in der integrierten Produkt- und Prozessplanung* (328 p.). Apprimus Verlag.

The Economist Intelligence Unit Limited. (2020). The great unwinding Covid-19 and the regionalisation of global supply chains. *The Economist Intelligence Unit*, 7 p.

Whelan, T., & Kronthal-Sacco, R. (2019). Research: Actually, consumers do buy sustainable products. *Harvard Business Review Online*. Accessed April 29, 2021, from https://hbr.org/2019/06/research-actually-consumers-do-buy-sustainable-products

Wiendahl, H.-P., Gerst, D., & Keunecke, L. (2004a). Variantenbeherrschung in der Montage. *Springer Verlag*. https://doi.org/10.1007/978-3-642-18947-0

Wiesendorfer, G. (2016). Instrumente und Prozesse zur Prognose des Absatzmarktes für Traktoren und Erntemaschinen aus Sicht des Branchenverbandes und der Industrie. Masterarbeit Hochschule Anhalt (FH).

Wildemann, H. (2020). *Variantenmanagement: Leitfaden zur Komplexitätsreduzierung, –beherrschung und -vermeidung in Produkt und Prozess.* Transfer Center for Production Logistics and Technology Management.

Winton, N. (2021). Global auto chip shortage, exacerbated by locked-down gamers, should recover soon. *Forbes Online*. Accessed February 14, 2021, from https://www.forbes.com/sites/neilwinton/2021/02/02/global-auto-chip-shortage-exacerbated-by-locked-down-gamers-should-recover-soon/?sh=6bbcd4784097

Womack, J. P., Jones, D. T., & Roos, D. (1990). *The machine that changed the world* (323 p.). Westview Press.

Basics on Takt and Flow: Insights from Practice

2

Everything should flow—evenly, transparently, and plannable. Nothing is more significant for the effective and efficient manufacture of products than the generation of a continuous flow of material, information, and work. In a customer-centric company with its diverse and varied products, this is a real challenge! In a fixed takt, a continuous flow of variant-rich products or even entire product portfolios is not possible. In the first sections of this chapter, we will demonstrate the importance of a continuous flow for production, in particular for assembly. We lay the foundation for being able to explain the relevance and capabilities of variable takt in the context of continuous-flow production later in this book. To do this, we combine theories of Lean Management with current scientific findings on flow production and supplement this with practical experience from the everyday life of a production manager. In doing so, we focus on balancing of an assembly line as it takes place in practice. We extend the classification of mixed-model assembly by three sub-types and separate product variance from option variance. Finally, we illustrate the occurrence of takt and model-mix losses, which we summarize as utilization losses. We show the first approaches to reduce these losses and will elaborate on them in the next chapters.

2.1 Benefits of Clocked Assembly in Continuous Flow

The origins of classic clocked assembly line work can be found historically both in the Scientific Management of Frederick Winslow Taylor, but also in the first practical applications of Henry Ford or Ransom Eli Olds (both American automobile pioneers). Since that time, the scientific community has dealt numerous times with the advantages of a clocked flow assembly. In most publications, the following advantages are highlighted and demonstrated (Scholl, 1999; Kratzsch, 2000; Boysen, 2005; Cachon & Terwiesch, 2020):

P. Bebersdorf, A. Huchzermeier, *Variable Takt Principle*, Management for Professionals, https://doi.org/10.1007/978-3-030-87170-3_2

- *High productivity*: no or little waiting time, high repetitiveness of activities
- *Good land use*: minimal area is occupied, as no buffers
- *Low daily operational control effort*: fixed, predetermined work schedule
- *Plannable quantities*: high availability, defined working and takt time
- *Short lead times, low capital investment*: no buffers, no finished goods inventories
- *Low labor costs*: low qualification requirements due to division of labor and standardization

Directly causally, there is nothing to add to these advantages. What these advantages of clocked flow assembly have in common is that they can be logically derived or calculated with the help of mathematical models. For this reason, the bulk of scientific research takes place in this area. It is driven by scientists whose achievement consists in representing the production world in highly stylized models with assumptions, deriving complex optimizations, and drawing conclusions on the basis of these. Extremely important and not to be underestimated results for a future-oriented production management are generated, i.e., decisive advantages for the management of a production can be drawn from scientific work. However, one important aspect is often missing in these models—the assembly of products is managed, planned, and performed by human individuals. These people do not always act logically, not always in the sense of the common cause (the enterprise) and certainly not always in the same way, despite the same basic conditions. However, it is precisely here that further managerial advantages can be drawn for companies through clocked flow assembly. From the approaches of Lean Management with thought leaders Taiichi Ohno (2013), James P. Womack et al. (1990), Jeffrey K. Liker (2004, 2021), and Mike Rother (2009) further advantages arise, which are of great importance for a successful production management. Many academic, but especially practical experiences show that visible methods and tools must not be copied, the underlying principles and philosophies must be understood and adapted to the respective company. For this reason, we mainly try to describe the idea behind variable takt, the "why." At the same time, we question the widespread assumptions of production research or even the Toyota production system and expand their solution space by the principles of variable takt, the "how."

One of the most important fundamentals in Lean Management is the pursuit of continuous flow. In the course of this book, we will be able to show that, on the one hand, **the assembly of different products with different total assembly efforts in a fixed takt cannot represent a continuous flow of workload.** When using variable takt, on the other hand, a steady flow of workload for all workers on the assembly line is established almost by itself.

2.1.1 Continuously Flowing Processes Bring Problems to the Surface (Liker, 2004, 2021)

Flowing processes lead to a "reduction in lead time from raw materials to the finished end product, outstanding quality, the lowest possible costs and the shortest possible

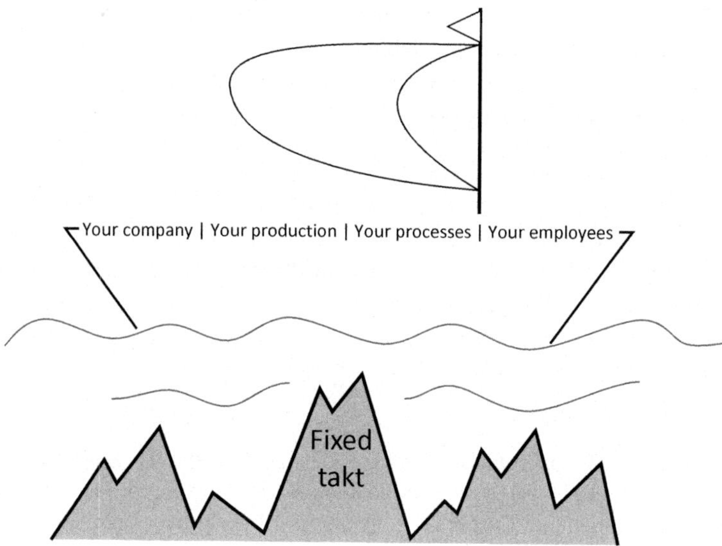

Fig. 2.1 Lean Management—problems must become visible

delivery times" (Liker, 2006, p. 136). How does Liker arrive at this? For every production manager, this translates into the proverbial "squaring of the circle" or in short, finding the optimum in the magic triangle of costs, quality, and delivery lead time.

Before answering this question, take a look at Fig. 2.1, which we are sure the reader is very familiar with.

A ship, your ship, is sailing over pointed and sharp-edged rocks; these rocks threaten to tear open the bow of your ship, your production, your company. Where is the connection to good management? Well, the rocks symbolize the classic problems in a production or company: poorly coordinated processes, lack of qualifications, unclear specifications and escalation levels, poorly maintained equipment, and much more. The water symbolizes the answers of poor management: stock levels and buffers, rejection stations in the production line, time buffers, and decouplings, to name just a few common answers in practice. The goal of effective management must always be to lower the water—that is how the rocks, and thus the problems, come to the surface and *must be* solved! To then circumnavigate the effects is no longer possible—good management must not make it possible. So, a rock could also stand for the existing line balancing of an assembly line with different products in a fixed takt—this can "tear apart" an assembly. Only with additional compromises in inventory or efficiency, these rocks in the fixed takt can be circumnavigated. If an assembly does not want to make these compromises, the problem of the fixed takt must be solved—variable takt would be one approach. Successful production management never opts for the short and fast route; it has been learned that these solutions produce worse results in the long term for the company, the customers, and the employees. The water thus symbolizes the "softeners" that cover up problems but do not force us to solve them.

Flowing processes bring cause and effect very close together in terms of time. A defective component, incorrect assembly, or an incorrect data record are detected immediately, as they cause a malfunction in the subsequent process and are not first collected in the buffer or warehouse. This not only avoids costs due to the continued production of defective parts, but the close temporal link between cause and effect also makes it much easier to analyze the root cause. The conditions that led to an incorrect assembly of the steering pressure pipe can be reconstructed much better after 20 min than after 2 days!

Production in flow dispenses with buffers and inventories, which extend every lead time and only increase the capital investment. Space is needed to (temporarily) store products—often one of the most valuable resources, especially in existing factories.

Since defects of various kinds develop their negative effects much more quickly in a clocked flow assembly, they force the organization to be disciplined, to eliminate defects on a sustained basis, and thus to continuously improve production processes and the organization.

In future matrix or even hybrid line assemblies (Sects. 2.2, 7.2, 8.1–2), it will be a success factor to maintain this "force to flow" of an assembly line and to create as few decouplings as possible to compensate for poor production management. We will show that variable takt can significantly expand the range of use of line assembly so that flowing processes can further bring weaknesses to the surface.

2.1.2 Overproduction Leads to Waste

Overproduction is also very often referred to as the "mother of all wastes" because if you produce more than you consume or your customers ordered, then you create (Sugimori et al., 1977):

1. *Inventories*: Too much produced goods occupy space and tie up capital.
2. *Transport*: Goods that cannot be shipped directly to the customer must be transported to intermediate storage facilities.
3. *Movements*: Too much produced goods have to be removed from storage, put into storage, repacked, or simply moved aside. This causes further costs and can lead to quality degradation or damage.
4. *Waiting times*: If too much has been produced and this overproduction has to be moved around the plant, there will be waiting times for other products that are currently needed.
5. *Rejects/rework*: If production is overproduced into an intermediate storage or buffer, then defective (intermediate) products are not noticed until much later in subsequent processes. The upstream process produces defective products for far too long.

6. *Double processing*: If there are no real customer orders yet, but production is carried out anyway, because "the machines have to run after all," products have to be processed a second time in order to finally adapt them according to the specific customer order, option variants respectively.

It seems trivial at first, on the one hand, a flow assembly minimizes (and in some cases even eliminates) these types of waste. This should be a flow assembly in total-mix or build-to-order. This refers to a production that, ideally, is precisely aligned with customer demand in terms of its production sequence and volume. If different products have to be assembled on one assembly line, this is a real challenge. Often, compulsory sequences or mix specifications have to be observed in order not to overload or underload the assembly line, resulting in overproduction. In extreme cases, products have to be built that are not in demand at all or can only be sold at a discount. Sometimes, the production volume for the products in demand can not be met. On the other hand, variable takt can help here, because when variable takt is used, a constant load is established regardless of the mix of products on the assembly line. All by itself a true build-to-order would be possible. Series-sequence adjustments due to the model-mix become unnecessary, overproduction is avoided.

2.1.3 Perfection as Aim

A continuously flowing assembly line forces perfection. Experienced foremen, team leaders, and assembly planners know it: the better you organize your line section, the less work, hassle or firefighting is necessary in the daily work.

Continuous flow assembly does not forgive mistakes and carelessness in employee qualification; insufficiently qualified workers produce poor quality and production losses. The answer is a standardized qualification of new employees in training centers, learning islands, or on the line. Existing employees must be continuously developed and their qualification level regularly checked.

If an assembly line is to be improved to perfection, it is a basic requirement to be able to stop the assembly line at every deviation from the standard. A detailed description of the Andon principle will be omitted here, this can be looked up in the numerous references to Lean Management (Liker, 2004, 2021; Ohno, 2013). Only by these stops the compulsion arises to solve the arising problems first immediately by means of short-term measures and later by problem solution processes lastingly.

An assembly line that starts at 6 a.m. every morning requires highly professional workforce planning. A clocked line with 100 workstations cannot start with 99 workers. If 101 workers are present, only 100 can still be productively deployed in the line. Precise attendance control with flexible compensation options for capacities must be established. It is important to ensure production capability at all times and still not lose sight of productivity.

Even if these constraints initially appear negative, they nevertheless force the organization and its management to continuously perfect existing processes and structures. Perfection is created in a continuous flow.

2.1.4 Line Pressure

One of the most critically discussed and ambivalent effects in practice is the line pressure perceived by all those involved (workers on the line *and* managers). This common term in practice and, above all, its effect should be familiar to anyone who has worked in or been operationally responsible for a clocked assembly line. The line flows relentlessly, it waits for no one and does not distinguish between beginners and "old hands." On the one hand, assembly line pressure leads to high productivity because it forces everyone to be perfect in leading, planning, and executing the assembly process. One mistake and the assembly line stops—this loss is irreversible and cannot be recovered later. A fact that, for example, does not strike in administrative processes in this severity, in this short term. Whereas delays in development processes are measured in days, in flow assembly, literally every minute, indeed often every second, counts. Truck drivers are called every 5 min or their position is called up every minute to find out whether they will reach loading ramp 1 at 12:30 p.m. or 12:50 p.m. In the case of a critical missing part, a difference that can account for 20 non-produced units and 1000 waiting workers. A continuous flow that does not compensate for mistakes forces all levels to perfection! Unfortunately, these effects are currently still given far too little attention in scientific work.

On the other hand, this line pressure must be handled responsibly; not every employee, whether worker or manager, is equally resilient or perceives this stress to the same extent. Working in a clocked assembly line is not only a physical but also a psychological stress for many employees. In the context of demographic change, this is a fact that more and more companies have to face. This is because the capacity of non-timed pre-assembly for the older part of the workforce, the classic solution, is limited and cannot be increased at will by insourcing (Loch et al., 2010).

2.1.5 Conclusion on Assemblies in Continuous Flow

From these examples of other benefits of using a flow assembly, i.e., a continuous flow of work, it can be seen that the benefits are not just in quantitative metrics such as productivity and capital investment. Perhaps the much larger and uncopyable benefit arises in management and employees in the "unleashing of creativity, continuous improvement, and the relentless pursuit of perfection" (Liker, 2006, p. 167).

However, scientific literature (Scholl, 1999; Kratzsch, 2000; Boysen, 2005; Pröpster, 2015; Cachon & Terwiesch, 2020) does not only point out the advantages of continuous flow assembly. The following disadvantages, perhaps better described as challenges, are mentioned. It should be noted that all authors implicitly assume a fixed takt:

- Low flexibility in the event of fluctuations in the number of units and mixes
- Takt compensation—unproductivity due to the use of the average takt for options
- Increased planning effort for different options
- Coordination effort with the works council or compliance with agreed-upon regulations in the event of changes in the assembly process
- Monotonous activities of the workers due to short-cycle work

Throughout this book, we will show that variable takt can significantly reduce some of these disadvantages and thereby increase the range of use of line assembly.

2.2 Structural Forms of Assembly

In this section, a brief overview of different organizational types of production will be given. The decision for one or a combination of different structural types is part of the planning of an assembly and thus of Assembly Design for Mixed-Model Assembly (AD for MMA, Sect. 1.4). The following overview is intended, on the one hand, to provide a theoretical basis for the use of variable takt and, on the other hand, to generate structure in the different designations of production and assembly systems. Frequently, no precise linguistic distinctions are made between different types of organization in practice, which often leads to two different types of organizations being discussed using the same designation. Or different terms are chosen for the same production organization type.

In the following, we use a combination of the structuring of Boysen (2005) and Günther and Tempelmeier (2012). In their work, they present production in a very structured way as a work system in which input objects are transformed into output objects by means of production factors (Boysen, 2005). We extend their representations by the terms hybrid and matrix assembly.

The spatial arrangement of labor and operating resources and the method of transporting the objects to be processed are chosen as the essential characteristics. If a production is organized according to its functions or technologies, then a job shop production is created. For a long time, fabrication plants, mostly suppliers for assemblies, were set up according to this type of organization. Thus, the edging benches were located in one corner, the welding workstations in another area, and as a rule, these areas were assigned to separate managers. In the meantime, this approach has become more than obsolete; today, production is organized much more according to the flow principle. As a result, fabrications are increasingly developing according to the principles and structures of an assembly line (see Sect. 2.3). The advantages of continuous flow described above also apply to fabrication. Today, fabrication areas are created in a U-shaped layout or in one-piece flow, previously pure domains of assembly.

In the object flow principle, the work systems and functions are oriented according to the flow of products. If there is no time constraint, the result is series production or series assembly. Unnecessary work steps can be skipped, but the flow cannot be left. The object to be processed always moves on to the next station when all work steps have been completed. Without a time-based link, large intermediate stocks and backlog situations arise very quickly in this type of organization.

If there is a time coupling, this can be due to the process; this is particularly prevalent in the process industry. This type is referred to as forced sequence production.

If the temporal coupling is dispositive, according to a plan specification, a takt is created and thus flow production or also called flow assembly. This flow assembly can be distinguished again with respect to its temporal coupling (Spur & Stöferle, 1986; Halubek, 2012). Thus, the object can be moved intermittently in a synchronously clocked manner. Intermittent clocking results in losses due to waiting times of the workers. Nevertheless, these organizational types of an assembly are still widely used, for example, it is applied in the i3 assembly of BMW in Leipzig (Germany), the MAN truck assembly in Munich (Germany), or in the assembly of the Audi e-Tron GT in Heilbronn (Germany). When a sound or light signal is given, all workers briefly leave the line, the assembly objects are moved forward by one station, and then assembly resumes on the stationary object. This temporal coupling can also be displayed asynchronously. The entire line does not move synchronously one station further (e.g., there is no technical prerequisite for this), the individual assembly objects are moved manually one station further in succession at a fixed takt, i.e., a gap thus moves against the direction of flow of the objects. The city bus and coach assembly of EvoBus (Daimler's bus production) assembles according to this system in its two plants in Mannheim and Neu-Ulm (both Germany).

If the transport of the assembly objects takes place in a continuously flowing manner (Fig. 2.2), then we can speak of continuous flow assembly. Most automotive assembly lines are organized according to this principle. This type of organization with continuous transport of the assembly object is one of the most important prerequisites for the use of variable takt, as we will derive in Chap. 4.

A further differentiation of flow assemblies is made in Sect. 2.5, because in a flow assembly it is possible to assemble in a batch as well as in a batch of size one, i.e., in a one-piece flow or in total-mix. When using the variable takt, a flow assembly can come a lot closer to a real build-to-order.

If production is organized according to the object flow principle (Fig. 2.2) and a unidirectional product flow is dispensed with, manufacturing and assembly islands are created. These can be fully automated. An example of this is Flexible Manufacturing Systems (FMS), which contain several machining centers and tools (Günther & Tempelmeier, 2012), the products are moved in these fully automated. It is further possible to bundle several manual assemblies or manufacturing processes in one island, the transport into and to the islands can be automated or manual. If several of these islands are combined, we can speak of matrix assembly, which is also referred to as Modular Assembly in some publications (see Sects. 7.2 and 8.1).

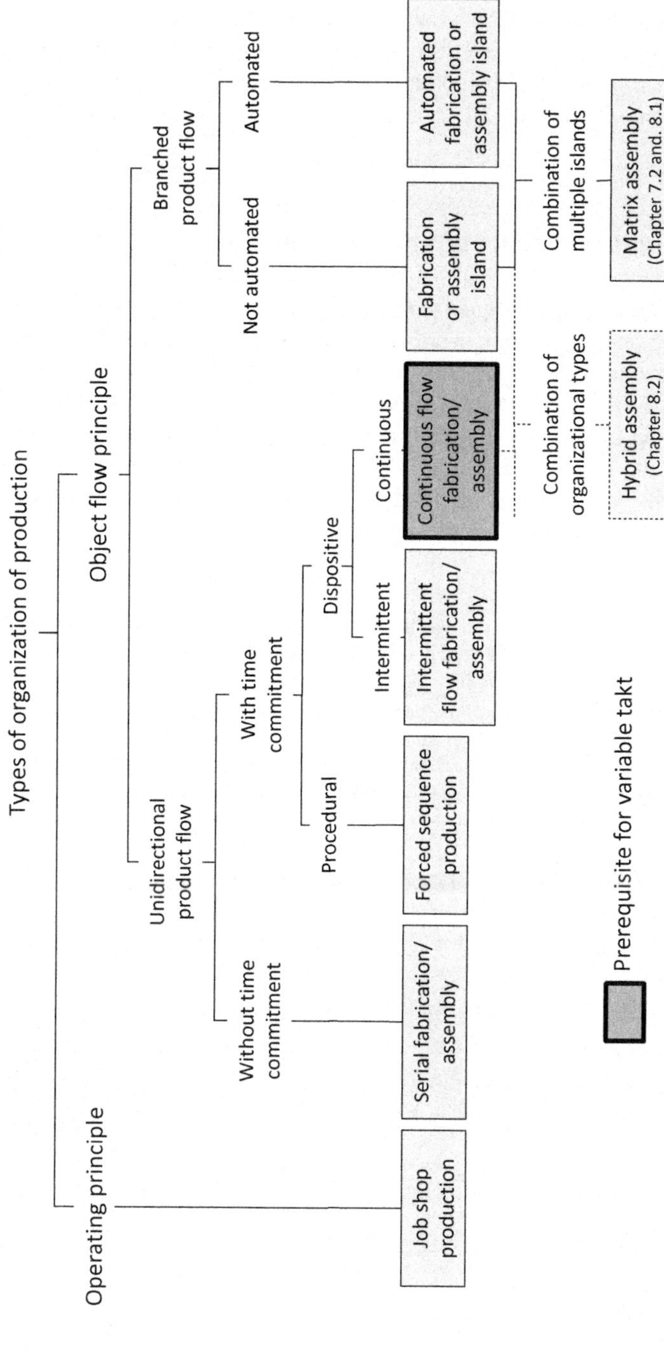

Fig. 2.2 Organizational types of production (adapted from Boysen, 2005; Günther & Tempelmeier, 2012)

Hybrid assemblies are not organizational types in their own right, they are created by combining at least two different organizational types. For example, a continuous flow assembly can be combined with an assembly island—a hybrid system is created. We describe a very interesting application of this combination in Sect. 8.2.

The production of complex products usually does not take place in only one type of organization; several types of organizations are combined with each other serially or in parallel.

2.3 Structural Forms of an Assembly Line

An assembly line, whether organized as a continuous flow assembly line or a series assembly line, can be set up according to different local structures. Each of these structures has its justification and its suitable area of application for different products or given infrastructural requirements. Provided that a continuous flowing transport of the products is implemented in these structures, an implementation of the variable takt is possible. The following figure shows the structures most commonly used in practice; other mixed types of these structures can and are used in practice. Figure 2.3 is based on the elaborations of Halubek (2012).

Open structures:

Line:	Is the structure that has found the most widespread use in practice. The biggest advantage is the very good accessibility for logistics and supply processes from both sides of the line and a low space requirement (Halubek, 2012).
U-shaped cell:	A modification of the line, with the advantage of a short return of the product carriers and good communication possibilities of the workers. The U-shaped cell is very popular in smaller assembly units where the worker can move from station to station with the product. In this way, the cell can be staffed with a variable number of workers to adjust the output of the line to the demand.

A practical example: A very successful implementation that combines the advantages of a line with those of a U-shaped cell structure is represented by the BMW plant in Leipzig. The assembly line and the building structure are designed

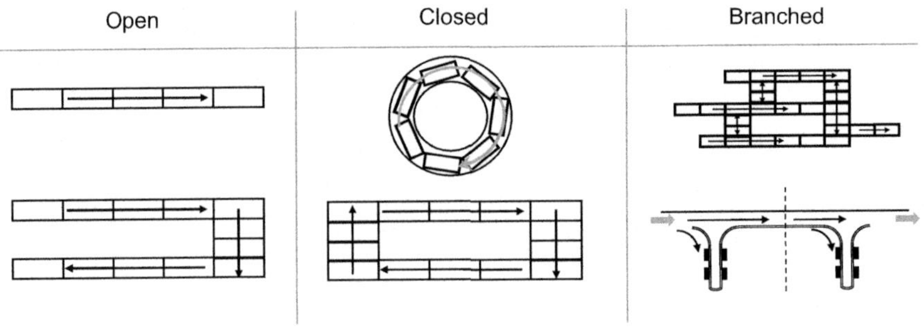

Fig. 2.3 Structural types of an assembly line (adapted from Halubek, 2012)

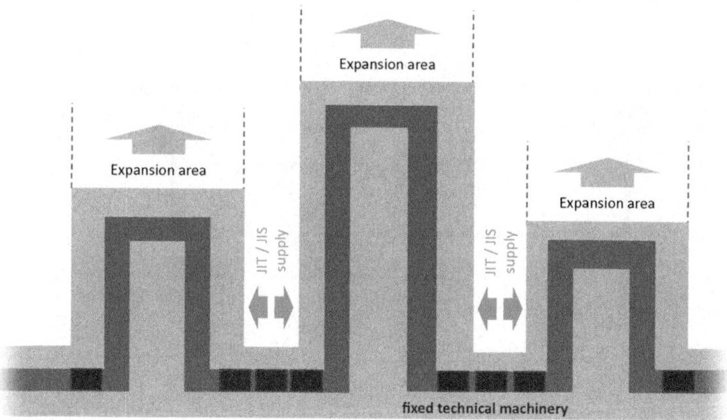

Fig. 2.4 "Finger structure" assembly line and building—own illustration adapted from the structure of the BMW plant in Leipzig (2019)

according to a "finger structure." Fixed technical and investment-intensive machinery are placed on the main line. The individual "fingers" contain mainly manual assembly activities. Since the building structure is adapted to the line structure, just-in-time deliveries can be delivered to the line by the shortest route and without stock transfers. Should an increase in capacity be necessary, the individual line sections in the "fingers," including the building cubature, can be extended—without having to relocate the fixed technical machinery, such as the wedding station or windscreen assembly (Fig. 2.4).

Closed structures:
Ring/Rectangle: These structures are very often used in pre-assemblies for a main line. The product carrier moves on an elongated rectangle in a circle, the interior space is thus usually reduced to a minimum. The material approach and processing can only be done from the outside.

Branched structures:
Can be implemented in the form of a network, optional auxiliary lines, or outfeed stations. An optional station selection is thus possible, variants can be processed on their own stations. A constant sequence cannot be maintained in this way, and the control effort also increases considerably. Modular assembly forms can be created in this way.

2.3.1 The Worker Triangle

In the further course, we will refer to the denotation of the worker triangle several times, but first a short explanation is required. Seen from a bird's eye view, the

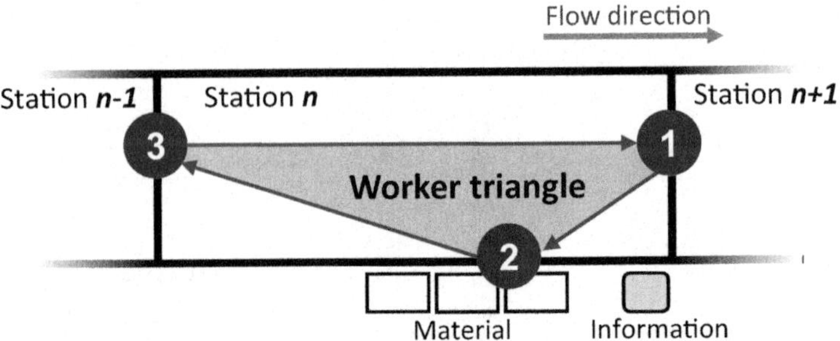

Fig. 2.5 The worker triangle

movement diagram of a worker in the assembly process resembles a triangle. Ideal for an assembly process with as little waste as possible are minimal paths (see Sect. 3.1 "Seven types of waste") in order to achieve a very high value-added share.

The worker always starts his/her new takt at the end of his/her station (1). The first thing he/she has to do is pick up the information about which product with which option variant is to be assembled next. For this reason, display systems such as MES (Manufacturing Execution System) screens, from which the worker can take this information, should also be installed at the end of the station. After the worker has been provided with this information, he/she walks back against the flow and collects his/her required material and any necessary operating supplies in the material zone (2). Having arrived at the start of the station, he/she starts his/her activity (3), flows down the line with the product and, in the ideal case, does not leave it again until the end of his/her station. A triangle has been created (Fig. 2.5).

For simplification, we assume in all our descriptions that only one worker w ($w = 1...W$) is used in a station n ($n = 1...N$). We are aware that in most assemblies, if the cubature of the product allows it, more than one worker per station s ($s = 1...S$) can be used. The ratio of all workers on the assembly line to the total number of stations is called worker density wd, $wd = (W/S)$. In order to minimize production area, and thus investment, an attempt is made to keep the worker density in a high range.

By the way, the ideal worker triangle is actually a point where the traveled paths of the worker have zero length, i.e., the triangle becomes a point. We describe this ideal approach in our chapter on Design-for-Takt, see Sect. 6.3.3.4.

2.3.2 Flexible Worker, Team Leader, Hancho

The original task of the flexible worker is to provide short-term support as an additional, broadly qualified worker in the event of problem cases or overloads at a workplace (Decker, 1993; Weiss, 2000). Flexible workers can be temporarily deployed from a pre-assembly area that is decoupled from the takt, or they can be deployed or held centrally via the assembly line. With the support of MES, flexible workers can be deployed proactively in advance of an overload, or they can provide support reactively when a need arises.

Despite a uniform designation, flexible workers sometimes perform very different tasks in the companies, such as:

- Balancing overloads or capacity peaks at individual stations
- Support of individual workers in case of disturbances in the assembly line
- Support or takeover rare options in the assembly line
- Replacement in case of short-term absence of workers
- Coverage of allowance times (takeover of a fixed number of takts for each worker of a group, for granting the personal allowance time of the worker)

As Lean Management and its associated tools and principles (Ohno, 2013; Liker, 2021) and improvement and leadership routines (Rother, 2009) have become more widespread, the function of the flexible worker has evolved in many companies. Team leaders are now used as an additional organizational level between employees and foremen, usually without disciplinary responsibility. A team leader usually leads a team of 8–15 employees. In some companies, these functions are referred to as Hancho, after their Japanese origin. "Han" in Japanese stands for "small group" and "Cho" for spokesperson. However, his/her area of responsibility goes far beyond that of a mere group spokesperson. The team leader or Hancho plays a key role in the continuous improvement of processes, and his/her tasks include:

- Reaction/support in case of problems—ensuring the assembly line runs
- Develop, monitor, and enforce standards
- Continuous improvement of processes

Some companies still define covering absenteeism and allowance time as part of a team leader's responsibilities.

2.4 Product, Option, and Variant: A Separate Classification for Variable Takt in Practice

The individualization of products has increased significantly in recent years and will become, or in many cases already is, a decisive competitive advantage for most manufacturing companies. Push systems of mass production have now given way to pull systems across the board (Fig. 2.6). This is not limited to single-product manufacturers but is transferring to a considerable extent to traditional mass manufacturers. The term mass customization (Pine et al., 1993; Kotha, 1995; Da Silveira et al., 2001) is used to summarize this development in literature and business. Shorter product life cycles and numerous new technical innovations reinforce this trend. Assembly, usually the last part of the manufacturing chain, has the greatest task in the individualization of products. With the aim of creating variants as late as possible, most variants are created in this area from the customer's point of view. An increasingly strong trend toward individualization through

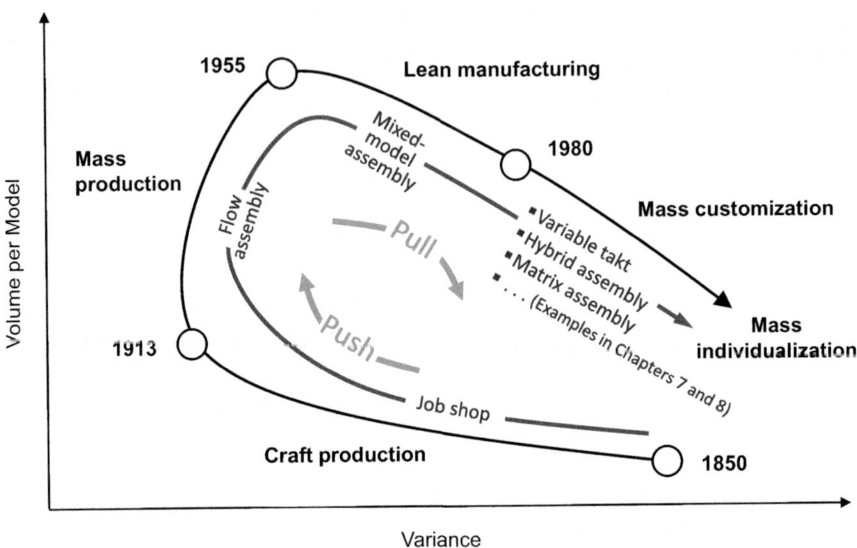

Fig. 2.6 Conversion of variance in manufacturing and assembly (adopted from Koren, 2010)—with addition of assembly structures

software based on hardware that is as identical as possible could replace assembly as the largest realizer of variants in the future, but we will not pursue this aspect further here. If this development is extended in thought and combined with current developments in the area of production technologies, this leads to mass individualization (Koren, 2021), as we already presented in Sect. 1.2. Koren (2010, 2021) depicts the transformation of variance in manufacturing and assembly very clearly in Fig. 2.6, which we have supplemented with the appropriate assembly systems. In the future, assembly lines will have to be able to accommodate even more products of different types, i.e., variable takt is one approach to minimizing the resulting efficiency losses.

First of all, we would like to bring a little more clarity to the term variant or variant diversity. A very good structuring of variant diversity in an assembly is presented by Wiendahl et al. (2004) in his work, which is highly recommended if the reader wants to delve deeper into the subject. The following considerations seem very theoretical at first glance. However, they are an elementary component for the understanding in which cases variable takt can be used, in which cases it unfolds its benefit and also the opposite.

Formally, according to the German standard DIN 199, objects of similar form or function, with a high proportion of identical parts that differ in one characteristic, are referred to as variants. Not only the DIN standard thus allows significant scope for classification, but also numerous scientific literature arrives at different concepts. Here we use an overview based on the work of Halubek (2012) and Swist (2014) (Fig. 2.7).

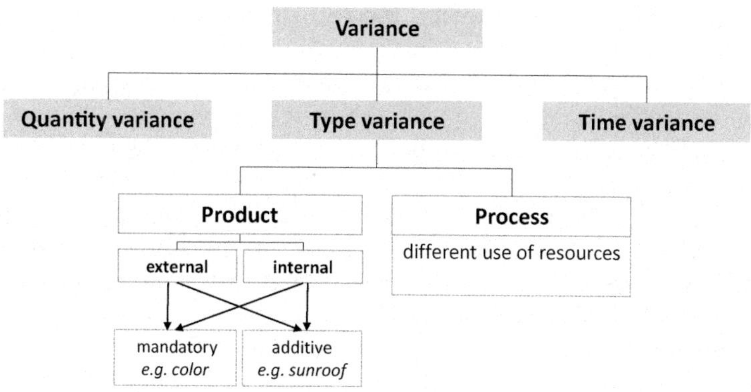

Fig. 2.7 Types of variance (adapted from Swist, 2014)

Quantity variance arises from a fluctuation in production output. In the entrepreneurial context, this is initially compensated for by changing the working hours and shift models. If this is not possible, the speed of the assembly line and thus the line balancing (Sect. 2.6) must be adjusted.

Time variance exists for goods that are only in demand during certain periods; these are usually seasonal products.

For our further considerations, the type variance is of decisive importance, since it is the main cause of process time spreads (Sect. 2.6.8). Type variance is divided into product variance and process variance. In the case of process variance, assembly resources are used differently (sequence of activities, technical parameters, etc.). Product variance is the differences in form, material, and function perceived by the customer. These can be additive, e.g., an additional luggage rack on a bicycle, or mandatory, e.g., the color of the bicycle. If the variance exists in the product, it can again be differentiated into an internal and external variance (Swist, 2014). The external type variance is the crucial one for the customer, because this is what he/she perceives. The most essential reasons for this external type variance are the customer's desire for individualization and adaptations to different markets (standards, legislation, approval conditions, etc.). Since internal type variance is not perceived by the customer and thus does not represent any value for him/her, it should be avoided as a matter of principle. Such internal variance often results from inefficient design processes and communication structures (Swist, 2014), for example, components are newly developed for which a validated version already exists.

Whereas in 1980, 100% volume growth was accompanied by 100% growth in the number of variants, in 2002 there were already 700% more variants with the same volume growth (Wildemann, 2020). Another 20 years later, in 2022, this value is likely to have increased many times over.

Later in this chapter, we will show that dealing with the variance in assembly times, which in turn arises from the type variance of the different options, is one of the biggest challenges in balancing an assembly line. Always with the aim of minimizing inefficiencies in the form of takt and model-mix losses (Sect. 2.8).

2.4.1 Delimitation of Product, Model, Order, and Option

Colloquially, the terms product and model are used very differently. The most common distinction is that the *product* can be divided into different *models.* We do not use this subdivision in the further course and set the designation *product* and *model* synonymously. We achieve no benefit in a differentiation of these two terms.

The terms *product* and *option* or *product variance* and *option variance* are used. An option represents a type variance characteristic of a product, this can be additive or obligatory. A product is an independent object that is also perceived by the customer as independent and clearly distinguishable in its basic structure from another product. In our definition, for example, the Audi A6 Avant is an option variant of the Audi A6 product. A Mercedes B-Class is in turn a different product, such as the Mercedes GLE or the E-Class. On closer inspection, even this classification attempt is not always clear-cut and can be perceived as blurred in many cases.

For this reason, we try to delimit both designations more deeply with further properties, always with reference to a later use in variable takt.

As already mentioned, the division of a company's entire product range into products, delimited by option variants, is usually not immediately clear (Table 2.1). However, the separation into different products must be made because different takt times can be assigned for different products on an assembly line with variable takt. This is not possible for different option variants, as we explain later (Sect. 2.5). The most striking distinguishing feature is the existence of independent standard worksheets (SWS) as we describe them in Sect. 2.7—each product needs an independent SWS, whereas different option variants are documented in one SWS.

Note: It is very common for products to be offered in or with option packages. An option package then combines a larger number of option items. This collection of options could lead to a situation where, for example, product P1 with the "Basic

Table 2.1 Subdivision of product and option variance

	Product variance – "P1 compared to P2"	Option variance – "A1 compared to A2"
Implicit categorization	• Is perceived as an independent product within the company	• Is perceived in the company as an option of a product
Structure design	• Significant structural differences	• Difference in terms of overall structure small
	• Own bill of material, independent design "tree"	• Part of a variant bill of material
Localization of variance	• Differences in assembly effort affect all stations **– Global variance**	• Differences of the assembly effort limited to a few stations **– Local variance**
	• Significant technical changes along the entire assembly line when new products are introduced	• In case of new option introduction local technical changes at individual stations
Structure assembly planning	• Line balancing documented on own standard worksheets (see Sect. 2.7)	• Only one option on a standard worksheet (see Sect. 2.7)

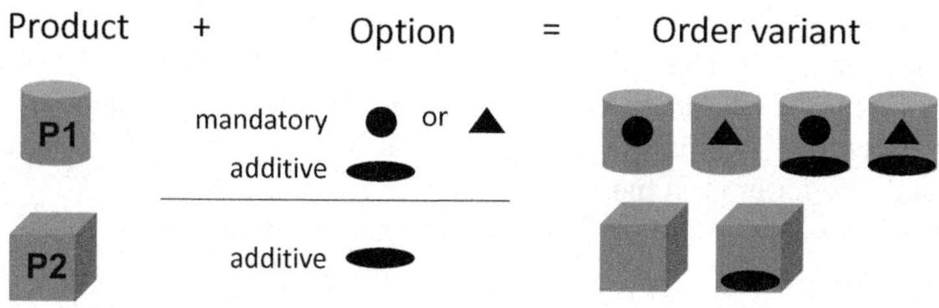

Fig. 2.8 Example for the creation of order variants

option package" differs significantly in structure and assembly workload from product P2 with the "Premium option package." In this case, it could be useful to treat these two option packages as independent products in order to minimize subsequent utilization losses (Sect. 2.8) by using variable takt. In this way, they could be given an independent takt time and an independent balancing.

We define an *order* as a specific individual customer order that is manufactured. If an assembly line produces 200 units per day, this also means 200 individualized orders—identical order variants are possible. The more variants the product has, the less likely it is that two exactly identical orders will be produced. In Fig. 2.8, we show an example of how two products and three option variants can result in a total of 6 order variants.

2.4.2 Local and Global Variance

If the customer selects a specific option for his/her order, the effect of this option in the assembly is usually limited to one or a small number of stations. For example, the additional loudspeakers and subwoofers of a sound system are installed at specific stations; this option has no effect on the remaining 95% of stations. For this reason, we refer to this variance, triggered by options, as *local variance.*

Global variance arises between different products. If two products differ significantly in their overall structure and thus also in their overall assembly content, these additional or reduced efforts or workload are distributed over the entire assembly line; in this case, we speak of *global variance.*

2.4.3 Choices in Dealing with Variance

Every manufacturing company should consciously deal with the variance of its products, because these decisions have far-reaching consequences for almost all areas of the company. Wiendahl et al. (2004) present four types of variance management in their work very well. We adapt the following representation around the increasing shift of variance into the software of products.

1. *Prevent variance*
   The greatest cost potentials arise from the fundamental avoidance of variance. The customer is offered no or only very few options. As described in the first chapters of this book, however, this is not in line with customers' increasingly strong desire for product individualization.
   Implementation level: *Product and company management*
2. *Reduce variance*
   Additive options could be included in the standard configuration of products. This is a step that must be well considered, because variance disappears, but also the potential to achieve additional margins.
   Implementation level: *Product and company management*

3.a *Shift variance—outside the manufacturing process*:
   Mandatory variants could be moved to the area of software, and the selection of the option is made by the customer him-/herself. For example, the customer of a passenger car can select different versions of his/her speedometer by means of a display. Mandatory or additive options can be generated by software and data sets, and thus neutrally for the manufacturing process. For many combustion engines, for example, the power class (rated power) is now set by means of a parameter set.
   Implementation level: *Product development*

3.b *Shift variance—within the manufacturing process:*
   Processes involving options can be transferred to other production areas (e.g., pre-assembly) or to suppliers (outsourcing).
   Implementation level: *Supply chain/production management*

4. *Mastering variance: The focus of this book*:
   Mastering variance is mainly performed in AD for MMA (Sect. 1.4) by methods of production design, planning, organization, and control. In SD for MMA variance is mastered by sequencing.
   Implementation level: *Production management*

2.5 Types of Mixed-Model Assembly

Based on the explanations in the previous sections on the structural types of an assembly and the distinctions between product and option, it is now necessary to delimit the term mixed-model assembly more precisely. A good summary on the classification of flow assemblies was made by Scholl (1999) and Boysen (2005), combined on their presentation and with our separation between product and option, we can classify the term mixed-model assembly in the following Fig. 2.9.

The simplest organizational type of an assembly line is single-product flow assembly (Günther & Tempelmeier, 2012). A uniform product without any option variance is produced. The utilization of this line can be kept very high without major effort. The only thing that has to be dealt with is fluctuations in the quantity demanded. If this is short-term, working times and shift models must be adjusted.

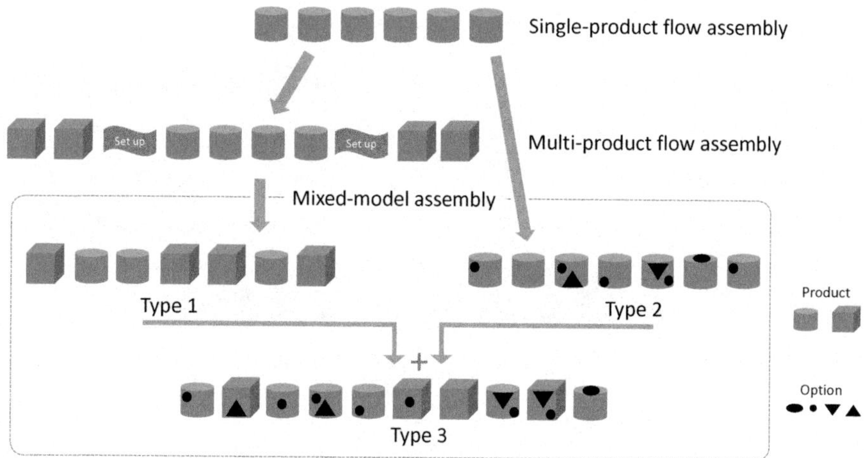

Fig. 2.9 Flow assembly systems; own illustration adapted from Scholl (1999), Boysen (2005) and Günther and Tempelmeier (2012)

In the case of longer-term fluctuations, the takt time must be adjusted. Variable takt is not necessary in this system.

In multi-product flow assembly (Boysen, 2005), different products are produced in batches. Between the batches, setup operations can, but do not have to, take place in the line. We deliberately use the term product rather than option variant in this case, narrowing the definition of Boysen (2005). In this type of organization, variable takt can be used in a goal-directed manner. Goal-directed refers to the goal of minimizing utilization losses (see Sect. 2.8).

Single-product and multi-product assembly lines can no longer cope with the increasing importance of mass customization and mass individualization (Sect. 2.4). Their flexibility is limited to a temporal and quantitative effect; they can no longer map increasing variability in the type variance of the products. To the contrary, mixed-model assemblies can implement different product variants very well (Fattahi & Salehi, 2009).

In mixed-model assembly, also known as variant flow assembly (Boysen, 2005) or variant flow production (Decker, 1993), different option variants *or* different products are produced in a mix on one assembly line. Set-up times must be reduced to a minimum in a mixed-model assembly, so batch size of one production is possible (Boysen et al., 2009a, b). It should be noted, however, that the increased use of mixed-model assembly is accompanied by increasing efficiency losses and control efforts (Moench et al., 2020a, b), which must be mastered and reduced. Internationally, the term mixed-model assembly line, in short MMAL, has gained acceptance (Thomopoulos, 1967). None of the work clearly addresses the distinction we have made between product variance and option variance. For this reason, we extend the structuring of mixed-model assemblies to include three subtypes. With this specification, we distinguish ourselves from current definitions in order to more clearly delineate the applicability and usebility of variable takt later in the book.

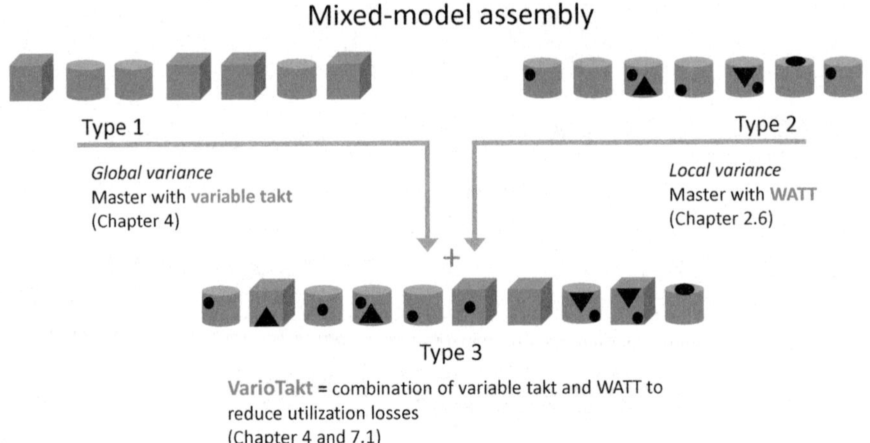

Fig. 2.10 VarioTakt in mixed-model assemblies of type 3

Note: At this point, we anticipate the further course of our work somewhat in order to reinforce the necessity of this differentiation once again and to provide an initial context for the topics of variable takt, WATT (weighted average takt time) and VarioTakt. We will explain the definition and composition of utilization losses in Sect. 2.8 (Fig. 2.10).

Mixed-model assembly of type 1:
Different products without option variance are produced in a mix. Variable takt can be used to increase utilization, respectively minimize utilization losses (Sect. 2.8). With variable takt, global variance is mastered. Derivation and justification are given in Chap. 4.

Mixed-model assembly of type 2:
A product in different variants (i.e., with different options) is produced in a mix. Since the variance is locally limited, variable takt cannot be used purposefully, but the WATT balancing method, which we present in Sect. 2.6, can. Combined with open station boundaries and drifting of workers, local variance is thus mastered.

Mixed-model assembly of type 3:
Is a combination and thus the addition of the type 1 and type 2. Different products with different option variants are assembled in a total-mix. Variable takt can be used purposefully to minimize utilization losses between products. In addition, the WATT takt method is used to minimize takt and model-mix losses (see Sect. 2.8) generated by option variants. We refer to the combined use of variable takt and WATT as *VarioTakt*. We go into a more detailed description and the origin of this designation in Chap. 4. With the VarioTakt local and global variance can be mastered, we describe a real practical example in Sect. 7.1.

The VarioTakt thus uses a combination of the fixed and variable takt. Utilization losses caused by variance in total workload between products are reduced with variable takt. Utilization losses caused by variance in workload between options in a product are reduced by using the WATT method in the fixed takt of the respective product.

2.6 Balancing: The WATT Method

An assembly line must be balanced when it is initialized and usually at regular intervals. This initial or reconfiguration of the flow system can have a variety of causes and can affect the entire assembly line or locally at individual stations (Boysen, 2005).

Possible causes are:

- The structure of the assembly line changes, the number of stations, technical equipment, etc.
- A change in the takt time is necessary, e.g., due to changes in demand quantities
- Change in demand of certain option variants (the "biggest evil" in automotive assembly, i.e., without volume change reconfiguration is necessary)
- New products must be integrated into the assembly line
- For the realization of productivity increases after reductions of assembly times of individual operations

Internationally, the terms "assembly line balancing" (Jackson, 1956; Scholl, 1999; Boysen et al., 2009a, b) or "line balancing" for short have become established. The theoretical basis of balancing has already been described in numerous scientific articles, but also in different textbooks. A very good overview is given by Günther and Tempelmeier (2012) in their textbook and Boysen (2005) in his dissertation. We would like to emphasize, and highly recommend for a more in-depth study, the work of Altemeier (2009) in particular, who describes very systematically the procedure for balancing of a variant flow assembly. Since the present work is intended to focus on variable takt, we will only go into the most important basics of fixed takt in the following. We will go into more detail on the WATT method, because we will combine this with variable takt to form the VarioTakt later on.

2.6.1 Input Variables

Before implementing a new line balancing, the following information should be available (Boysen, 2005; Altemeier, 2009):

- Forecasted demand quantity of each product
- Precedence graph of each product
- Restrictions (e.g., facilities, equipment, personnel)

- Desired working time model and derived target takt time (number of workers W results in the line balancing)

or

- Number of existing workers W and desired working time model (target takt time results in the line balancing)

To simplify matters, we will assume in the remainder of this chapter that the takt time is given. In Sects. 4.4.3 and 4.4.5, we will work out the derivation of the appropriate variable takt time for a given product portfolio in a fixed takt, which is to be converted to a variable takt.

2.6.2 Work Operations

In the first step of line balancing, the product or production task is broken down into all its work steps, or operations respectively. These work operations are "a set of work elements that can be performed on an object in a spatially and temporally self-contained manner" (Domschke et al., 1997). In the context of assembly planning, these work steps are also referred to work operations and abbreviated as WO; we will use this abbreviation in the further course. The necessary level of detail must be determined independently and usually depends on the targeted takt time and thus indirectly on the volume to be produced. In practice, for example, the WOs are not broken down in as much detail for a targeted takt time of 70 min (approx. 2000 units of annual production for a one-shift operation) as they are for a targeted takt time of 60 s (approx. 200,000 units of annual production, for a two-shift operation). The shorter takt times make smaller WOs necessary; this increased administrative planning effort is usually justified for large quantities by improved productivity (economies of scale due to lower takt losses with small WOs, see also Sect. 2.8.1). Safety-relevant assembly activities, such as the braking system of a vehicle, or legal requirements, as in the production of aircraft, may force additional detailing of the work operations. Ultimately, this decision remains a trade-off between benefit and expense and must therefore be determined by management.

2.6.3 Operation Times and Performance Level

In the next step, operation times with a target performance level are to be assigned to the work operations. REFA and MTM methods (time and motion studies) or a variety of other procedures can be used to determine the operation times. Due to different legal or collective bargaining requirements, a large number of different procedures and regulations are in use internationally. In Germany, for example, the use of these methods must be agreed with the employee representatives, e.g., the works council.

"The performance level is an assessed percentage addition or subtraction to a human work performance measured in time and is used to normalize individual performance characteristics when transferred to 'collectives'" (Schlick et al., 2010). This performance level is assessed by a trained professional and is always related to an expected performance level of 100%. Usually, the target performance levels of the workers on the assembly line are set to the same value. In a clocked assembly line, a worker has no influence on his/her personal performance level; this is set for him/her by the workload combined with the line speed. If the current line situation allows it, the worker can selectively adjust his/her individual performance level—positively or negatively, he/she then works ahead or falls behind. We will not go any further into the determination of operation times and the performance level.

2.6.4 Precedence Graph

Once the production task has been broken down into its individual work operations and operation times have been defined, a precedence graph is created in theory (Fig. 2.11). A product to be assembled contains technically determined restrictions during assembly (Decker, 1993; Altemeier, 2009). This results in precedence relationships, or in other words, there are activities that must necessarily be performed before other activities. For example, the assembly of the wheel bolds must take place before the assembly of the wheel (Fig. 2.11).

If several products are assembled in an assembly line with a fixed takt, in most cases an attempt is made to create a common precedence graph for all products. The higher the number of different products on the assembly line and the higher their option variance, the more complex this precedence graph becomes. When using variable takt, it has proven useful to dispense with a common precedence graph. Each product is balanced independently and the WOs are documented on separate standard worksheets (SWS, Sect. 2.7) for each product and worker. It goes without saying that technical restrictions due to operating equipment that is only available at one station must nevertheless be observed. Ultimately, it is up to management to decide how strongly these restrictions should be set, or whether it makes sense in some cases to duplicate equipment. For example, it may be the case that the same

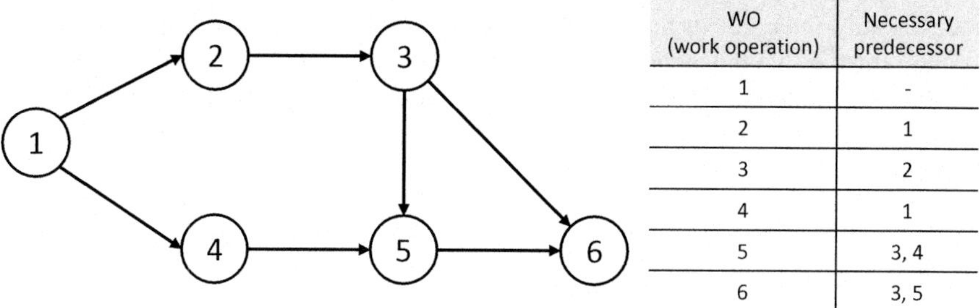

WO (work operation)	Necessary predecessor
1	-
2	1
3	2
4	1
5	3, 4
6	3, 5

Fig. 2.11 Simplified example of a precedence graph with six work operations

activity for two products can take place at different stations. The material provision for this identical WO would have to take place at two stations. A result which at first appears to be unfavorable. This can be circumvented by switching to (attached) traveling kitting carts for material staging. A traveling kitting cart contains all materials of an order for a certain number of stations. At which station the material is then removed is irrelevant. Due to the fundamentally increased space requirements for material provision on the assembly line in mixed-model assembly (especially in type 3), traveling kitting carts will become increasingly established.

Almost all methods based in science require a complete precedence graph. A complete precedence graph is also the prerequisite for an intended fully automated balancing by a software. In our experience, three critical observations arise in practical applications:

1. Even for an averagely complex product, the creation of a complete precedence graph represents an extreme effort. For example, Sternatz (2010) states that the effort required to create a complete precedence graph for a product with 15,000 WOs results in an effort of two man-years. In practice, we are not aware of any company that has a complete precedence graph for the assembly of a complex product. Currently, first approaches of learning, but not complete, precedence graphs are introduced in cooperation between science and practice at the car manufacturer Volkswagen. These are based on the work of Klindworth et al. (2012) and Otto and Otto (2014).
2. It is assumed that the operation time of a WO always keeps the same value, regardless of the selected assembly sequence. In practice, this is not the case. For example, it is true that WOs 2 and 4 (see Fig. 2.11) can be assembled in any order. However, it is possible that step 4 will somewhat hinder step 2 and thus the assembly time of step 2 will be somewhat extended if step 4 is performed beforehand.
3. The assembly times are assumed to be unchangeable. If it should turn out during a line balancing that, for example, the operation time of WO 5 would have to be reduced somewhat in order to significantly reduce utilization losses, then in business practice an attempt would be made to reduce the operation time of WO 5 by means of improvement workshops.

Unfortunately, line balancing with an incomplete precedence graph hardly receives any attention in the scientific literature. Thus, the most important source of data for the line balancing of an assembly line remains the (experience) knowledge of the assembly planners and workers on the assembly line. Distributed among a large number of these individuals, a complete precedence graph is at least implicitly available.

2.6.5 Objectives of Line Balancing

While the goals of line balancing can vary widely (Boysen et al., 2007), i.e.:

- Increasing line efficiency by reducing takt losses
- Minimization of takt time or maximization of production output
- Minimization of the number of stations or shortening of the line length
- Minimization of the lead time
- Levelling of station times

However, it can be assumed that the predominant goal of the vast majority of line balancing is to achieve the highest possible productivity (Altemeier, 2009) by maximizing the balancing efficiency (Sect. 2.6.11). In the case of extremely high product complexity, the first step may be to create a feasible line balance in the first place (Boysen et al., 2007).

2.6.6 Scientific Approaches to Line Balancing: The SALBP and GALBP Methods

A great deal of scientific work has been done in the last decades in the systematic solution of balancing problems. Since the first mathematical elaboration by Salveson (1955), research has essentially been concerned with the goal of distributing the work operations as optimally as possible, with different target definitions, among the stations (Boysen, 2005). Thus, a first branch of research was born after Baybars (1986)—the Simple Line Balancing Problem (SALB) method. Since it was quickly recognized that the premises of the SALB were too narrow for practical applications (Boysen, 2005), further work was combined into the General Assembly Line Balancing Problem (GLABP) method (Baybars, 1986). A good overview of the SALBP and GALBP methods is given in the works of Shtub and Dar-El (1987), Scholl (1999), Boysen (2005), Boysen et al. (2007), and Günther and Tempelmeier (2012).

2.6.7 Line Balancing in the Entrepreneurial Practice

We will describe line balancing from the user's point of view in business practice. We explain the WATT (weighted average takt time) method in detail. To simplify matters, the takt time is an input variable in the following descriptions and is therefore given.

In day-to-day business, line balancing is carried out by assembly planners in cooperation with the team leaders and workers of the assembly section concerned. It is carried out on the basis of empirical knowledge or short assembly tests. This trial procedure (Pröpster, 2015) does not lead to the ideal result, but algorithmic procedures are often not possible due to the lack of a fully comprehensive database and precedence graphs.

Since a new line balancing always represents a greater planning effort and can lead to training costs and losses, an attempt should be made to reduce a new line balancing to a minimum. Every plant manager or assembly line manager first tries to avoid a new line balancing by other means, such as changing the work schedule or adapting operating resources. In practice, it is known that a new line balance is not only associated with considerable administrative costs and costs for training the workers. A new line balancing always represents a special challenge to maintain the achieved quality level of an assembly line. In most cases, new line balancing is accompanied by increased error rates. In Sect. 4.5, we will show how the need for additional line balancing can be significantly reduced by using a variable takt.

2.6.8 Procedure of the WATT Method

First, the probabilities of occurrence P_l are defined for all WOs l that represent an option. This can be done on the basis of historical data or on the basis of a forecast for the future planning period. The time intervals at which these probabilities of occurrence are to be determined must be defined. If the probabilities of occurrence of the options in the production period deviate strongly from the planning phase, this can lead to increased model-mix losses (Sect. 2.8.2) due to increased over- or under-utilization of the workers. In addition, in many countries, these rules have to be coordinated with employee representatives. Often, the need for new line balancing arises purely from changes in the probability of options or from changes in the model-mix.

In most practical tasks, the takt time is specified in order to be able to produce a certain volume in a defined working time. In these cases, the WOs are assigned to the stations until their weighted sum reaches the takt time, but does not exceed it. To obtain the weighted operation time relevant for line balancing, the operation time ot_l of the WO is multiplied by the probability of occurrence of the option P_l. Figure 2.13 shows an example of the distribution of 10 WOs over 3 stations. In the following considerations, we always assume that exactly one worker is assigned to a station, i.e., $N = W$ (Table 2.2).

The sum of the weighted operation times ot_l at station n, gives the weighted station time R_n (following Witte, 2007; Swist, 2014).

$$R_n = \sum_{l \epsilon L_n} (ot_l * P_l)$$

with

R_n Weighted station time at station n
l Index of assembly operations
L_n Index set of assembly operations assigned to station n
ot_l Operation time of the assembly process l
P_l Probability of occurrence of the assembly operation l

This example can be displayed in a T-bar diagram. The designation T-bar results from the selected display type in the form of a bar for the average value and the maximum and the minimum station time in the form of a "T."

Table 2.2 Example of line balancing for three workstations

Station 1				Station 2				Station 3			
	ot_l	P_l	$ot_l * P_l$		ot_l	P_l	$ot_l * P_l$		ot_l	P_l	$ot_l * P_l$
WO 1	1	1	1	WO 5	4	1	4	WO 8	6	1	6
WO 2	4	1	4	WO 6	5	1	5	WO 9	2	1	2
WO 3	2	1	2	WO 7	2	1	2	WO 10	6	1/3	2
WO 4	4	1/2	2								

2.6.9 Limitation of Overtakting

In our previous example (Fig. 2.12), a maximum overtakting of 3 time units (TUs) occurs at station 3 (14 TU maximum station time—11 TU takt time = 3 TU). In order to avoid or reduce later model-mix losses specifically (see Sect. 2.8.2), it is recommended to limit overtakting specifically with measures. Usually, the amount of the maximum permissible overtakting has to be agreed upon with the employee representation in many companies. In our previous example of balancing, an additional WO 11 could be included at station 3 with an operation time ot_l of 10 TU and an occurrence probability P_l of 0.1. The weighted operation time would be only 1 TU and the weighted station time R_3 would be exactly on the takt time of 11 TU (Fig. 2.15).

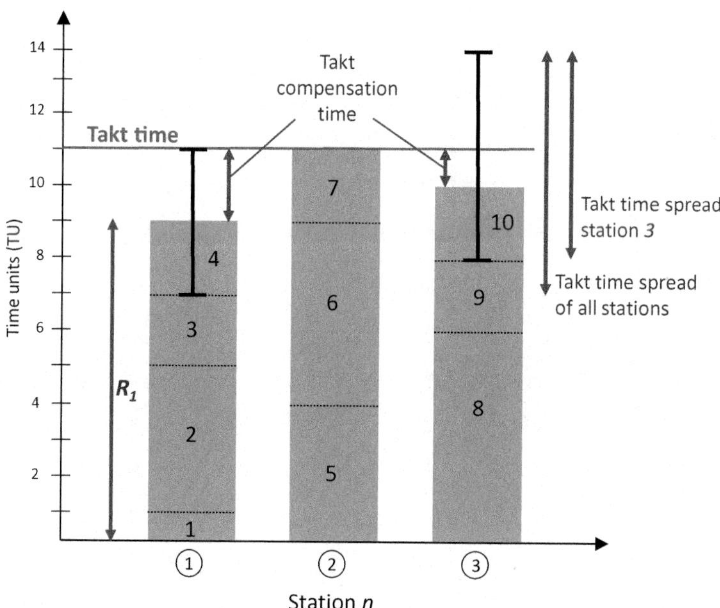

Fig. 2.12 T-bar diagram for line balancing of three workstations

However, in this situation, the practitioner immediately recognizes that an order containing the option of WOs 10 and 11 would burden the worker with 24 TU and this with a takt time of 11 TU. Measures to compensate for this overload would be very far-reaching and costly. In practice, it is said that large time modules with a low probability of occurrence (complex but rare option variants) "destroy" the line balancing. In a study of a premium car manufacturer in cooperation with the Karlsruhe Institute of Technology (Gans et al., 2011) it was shown that already 10% of the orders drift to 50% into the next station—a pure compensation via open station boundaries is thus not possible. Overtakting of option variants, and thus drifting, is also limited by the range of necessary operating resources. For this reason, it is advisable to limit the maximum overtakting already during balancing. The value at which the limit should lie depends on several factors, which we explain in Sect. 2.8.2 (Fig. 2.13).

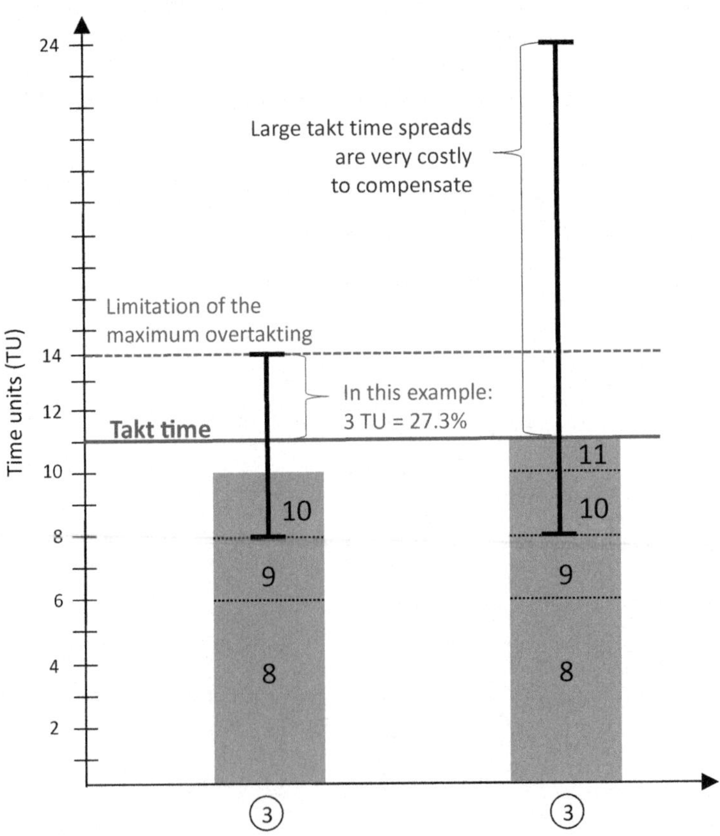

Fig. 2.13 Limitation of overtakting through balancing at workstation 3 with additional work operation 11

2.6.10 Open and Closed Station Boundaries

We would like to define the terms station or station boundaries and make a distinction between open and closed station boundaries. In our interpretation, a station is the smallest locational unit or the smallest layout element of an assembly line. In the best case, aiming at a minimum length of the whole assembly line, the stations are directly adjacent to each other and their addition gives the total length of the assembly line. In the fixed takt, the relation between the station length and the speed of the assembly object is constant. As a rule, the station length of all stations of a line is identical and so is the conveying speed. If the station lengths are not identical (e.g., technically necessary on a line), the speed of the assembly object must be changed in this station so that the takt time remains constant. In short: If the station n_x is 20% longer than all other stations, then the speed V_{nx} in this station must be increased by 20% so that the station n_x is left exactly after the takt time T. We thus set the goal or assumption that the station length should always be equal to the takt time. This is not mandatory, but in our following descriptions we assume it.

Depending on the size of the assembly object, one or more workers can work at a station. Since the station length should correspond to the takt time and overtakting is necessary in the WATT method, the worker leaves the station for products where the variant-dependent assembly time at the station is greater than the takt time—he/she drifts into the next station. Such a station boundary is called open. If he/she is not allowed to leave his/her station boundary, it is a closed station boundary—the assembly line must be stopped or an additional worker (a flexible worker) must take over his/her remaining work when the station boundary is reached.

Open station boundaries must be limited—an infinitely open station exists in some theoretical elaborations, but not in practice. Either for technical or organizational reasons, every open station boundary must be limited. In short: either the technical system has a limited range or the worker drifts so much that he/she disturbs other workers in their activity or prevents their activity, at the latest then the assembly system has to be stopped. In assembly lines with open station boundary limits, this limit is called the final station boundary. The final station boundary can be technically determined, e.g., the effective range of a piece of equipment, or it can be fixed (Boysen, 2005). At these boundaries, in and against the flow direction, the model-mix losses from Sect. 2.8.2 occur. The range of final station boundaries specifies the possible drift range of the station (Fig. 2.14).

A note at this point: With a deeper understanding of the variable takt process, it becomes apparent that a "fixed" station represents an artificial restriction. It is not in itself necessary or even feasible when using variable takt. In Sect. 4.4.4, we describe the view of the "virtual station lengths" that arise when variable takt is used. For the orientation of the workers, there will still be color-coded boundary lines as station boundaries on the assembly floor. When using the variable takt time, however, this one line is only valid for one of the takt time groups.

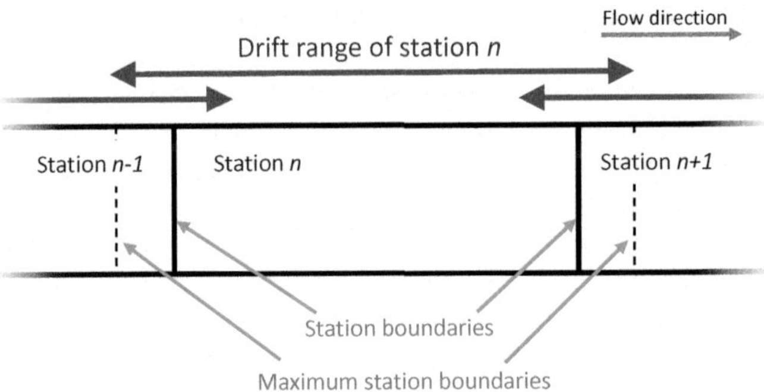

Fig. 2.14 Visualization of station boundaries

2.6.11 Local and Global Takt Time Spread: Important Key Figures of the Line Balance

If the takt time is not given, but the number of stations or workers, the ideal takt time should first be calculated.

$$Ideal\ takt\ time = \frac{Workload}{Number\ of\ workers}$$

Admittedly, this formula is oversimplified. On the one hand, it would correspond to a balancing efficiency of the assembly line of 100%, which would never be feasible in practical applications. Work operations cannot be divided arbitrarily small to fill all T-bar diagrams to 100%. In Sect. 4.4.3 "Determining the variable takt time," we explain in detail how a takt time is to be determined taking into account allowance times and other losses, for this reason we will not go into this in more detail here.

As a basic condition, it must be ensured that each weighted station time R_n is less than or equal to the takt time T (see also Günther & Tempelmeier, 2012). Furthermore, attention should be paid to a fundamentally technically feasible sequence and to existing restrictions, such as the position of the operating resources or employee qualifications. Restrictions could also result from the provision of materials. For a given takt time, all WOs must be distributed among the workers. By minimizing the number of workers required, maximum balancing efficiency of the assembly line and thus maximum technical productivity can be achieved.

Maximizing productivity while complying with all restrictions and quality requirements is the predominant goal of the vast majority of line balancing. In combination with minimizing the investment required for a new assembly line, the focus is always on reducing manufacturing costs.

One of the most important levers for maximizing the productivity of an assembly line is maximizing *balancing efficiency of the assembly line.*

Balancing efficiency
The ratio of the accumulated weighted station times R_n (in our example in Fig. 2.13 = 30 TU; one worker per station) and the maximum capacity (3 × takt time T = 33 TU) is called the *balancing efficiency*. In our example from Fig. 2.13, the *balancing efficiency* is 91%. Thus, the balancing efficiency describes the technical productivity of the assembly line (Swist, 2014). The balancing efficiency is often also referred to as the balancing rate.

$$E_{\mathrm{AL}} = \frac{C_{\mathrm{b}}}{C_{\mathrm{v}}} \cdot 100[\%] = \frac{\sum_{n=1}^{N} R_n}{N \cdot T} \cdot 100[\%]$$

with

E_{AL} Balancing efficiency of assembly line
C_{b} Actual capacity required
C_{v} Provided capacity
n Index stations $n = 1 \ldots N$
R_n Weighted station time at station n
T Takt time

The reciprocal of the balancing efficiency gives the takt compensation (rate).

Takt compensation
The time difference between the station time and the takt time is called takt compensation time or, in relation to the total capacity supply, takt compensation rate (see also Fig. 2.12). In our example at station 1, the takt compensation time is 2 TUs; for our entire assembly line with three stations, it is a cumulative 3 TU. This results in a takt compensation rate of 9% (cf. Koehter, 1986).

$$TA = \frac{N \cdot T - \sum_{n=1}^{N} R_n}{N \cdot T} \cdot 100[\%]$$

with

TA Takt compensation rate

Local and global takt time spread
The time difference between the lowest and the highest station time, caused by one or more options, is called takt time spread. In our example in Fig. 2.12, this is 14–8 = 6 TU at station 3. Takt time spreads can be specified for a station or the entire assembly line (see Fig. 2.12).

Local and global takt time spreads arise from local and global variance, as we describe in Sect. 2.4.2.

Option variants usually generate ***local*** takt time spreads; these takt time spreads are limited to one or few stations. For example, the components of option A1 are only installed at a few defined stations, the rest of the assembly line remains unaffected.

Local takt time spreads can be mastered via:

- Pre-assemblies (Sect. 7.1.2)
- WATT balancing (Sect. 2.6)
- Flexible workers (Sect. 7.1.3) or
- Hybrid assembly structures (Sects. 7.1 and 8.2)

The structural workload differences of global variance between different products generate most *global* takt time spreads. In other words, the costly product P1 has an additional cost distributed over the complete assembly process.

- It is precisely these differences in the global distribution of the assembly effort that can be mastered with variable takt (Chap. 4). This is exactly where variable takt shows its great advantages!

2.7 Standard Work and Standard Worksheets

Standard worksheets (SWS) are used to document and visualize the line balancing. The variants, forms of application, and technical aids for SWS are now very diverse. In the following, we will only go into the basics so that we can use the term SWS in the later course of this work.

The manifestations of SWS vary widely, but should essentially contain at least the elements we show in the example in Fig. 2.15.

Although the initial task of an SWS is to document the work steps and their sequence, in the context of Lean Management, the SWS has a further, very important meaning. Fujimoto (1999), Ohno (2013), and Liker (2021) describe this in great detail in their work. The SWS documents the agreed standard, the current best-known sequence of assembly work (Ohno, 2013). It is therefore the prerequisite for ensuring the level of quality and the basis of the continuous improvement process.

Quality assurance The SWS makes a very important contribution to ensuring assembly quality, as it describes in detail how the frontline employee works and not just the final state of his/her activity. For example, a final assembled flange may not differ visually if the bolting sequence is different, but the stresses in the component are different. Systematic troubleshooting in the event of leaks is then extremely difficult.

Worker: **MT_012_L_W1** Product: **T954** Takt time: **1.9 min**

Sequence	Identification number work operation (WO)	Description	Details	Option	Part number	Operation time ot_l	Probability of occurrence P_l	Weighted operation time $R_n = ot_l{*}P_l$	Notes/ pictures, etc.
1	003-A05-0023	Assembly step	Fetching the assembly step from the staging area and docking it to the AGV.			0.3	1	0.3	
2	003-B03-0001	Pre-assembly bracket	Preassemble bracket with two rubber buffers.		A520.6235.84 X254.698.58	0.25	1	0.25	
3	003-B03-0002	Mounting bracket	Mount bracket with 3x6 screw. Observe screw sequence.		X365.874.58	0.8	1	0.8	Visualization screw sequence
4	003-C55-0001	Fix harness	Fixing harness X325 with two cable ties.	V036	X658.987.54	0.3	0,5	0.15	Visualization fixation points cable ties
5	003-V01-0002	Hydraulic muff torque	Tighten the muff of the power steering pump with 33 Nm.			0.33	1	0.33	
							Total R_n:	**1.83 min**	

Fig. 2.15 Highly simplified example of the content of a standard worksheet (SWS)

Prerequisite for the continuous improvement process If the goal is to continuously reduce assembly times and thus reduce manufacturing costs, waste in the process must be identified and eliminated. But how can the search for waste be started systematically if the work process is different for all workers? Standard work is seen in Lean Management as a prerequisite for any improvement (Liker, 2021).

In addition to documenting standardized work processes, the SWS is used in day-to-day assembly operations primarily for training new employees and performing daily standard work process confirmations.

Standard work process confirmations Are the most important element for continuous improvement of an assembly line! The team leader carries out several process confirmations per day, in which he/she observes a takt of an employee on the basis of the SWS and tries to determine deviations. The deviations are to be questioned in the next step together with the employee concerned. If they represent an improvement of the process, the SWS is to be adjusted. Otherwise, the employee must be instructed again. In contrast to a technical machine, the ingenious thing about humans is that they improve themselves and their activities to a certain degree. In addition to the deviations, further improvement ideas will automatically "catch the eye" of the team leader, who can implement them immediately or document them first. The process confirmations by means of SWS represent the practical application of the Deming cycle or PDCA cycle (Plan-Do-Check-Act). According to Deming (1982), continuous improvement of any process (management processes and assembly processes) occurs when one plans the process (Plan), executes it (Do), then checks for deviations (Check), and reacts to deviations with improvement measures (Act).

As a rule, standard work process confirmations are carried out in the form of a cascade from the team leader to the plant manager. Whereby the next higher process always observes and improves its upstream process. The team leader thus observes the assembly process, and the foreman in turn observes the execution of the team leader's process confirmation. The foreman thus does not improve the production process, but the process of the team leader's process confirmation, and so on—a coaching cascade emerges. Since Mike Rother (2009, 2013) described this scientific approach of an improvement and coaching kata in his book "Toyota Kata," this approach has found practical application worldwide.

2.8 Utilization Losses

In Sect. 2.4, we showed the four basic approaches to counter variance in the product spectrum. If it is not possible or desired to prevent or reduce variance or to shift it outside the assembly line, then it must be mastered on the assembly line. Variable takt or the use of the WATT method are just two possible solutions. Whichever path you take in your assembly line, it will unfortunately inevitably lead to utilization losses. However, you can influence the extent of these losses.

In an ideal single-product flow assembly without any variance, it is possible to occupy every minute of workers' presence with productive work. In this sense, work is productive if it produces the planned value added in the planned time. What may already prove to be extremely complex in a single-product flow assembly will be almost impossible in a type 3 mixed-model assembly. There will be utilization losses. These utilization losses occur whenever the capacity available from the worker cannot be filled with productive work. This loss becomes visible through:

(a) Waiting times of one or more workers

or

(b) Unproductive repairs and reassemblies

Utilization losses can be divided into two types, takt losses and model-mix losses (Koehter, 1986).

$$Utilization\ losses = Takt\ losses + Model\text{-}mix\ losses$$

Koehter (1986) uses the term "model sequence costs" in his paper; we deliberately use the term utilization losses for them here because we believe it better reflects the nature of these losses. We retain this separation into takt losses and model-mix losses.

2.8.1 Takt Losses

Takt losses occur in the balancing phase when planning the assembly line. They correspond to the takt compensation that occurs during line balancing (Koehter, 1986). It is an average value that is valid for an unlimited period of time as long as the existing line balancing is used and the probabilities of occurrence of the option variants apply. Takt losses always occur in the form of waiting times of the workers. In the further course of this book, we will show that takt losses can be significantly reduced by using variable takt in a mixed-model assembly of type 1 or 3 (Sect. 4.5).

Figure 2.16 shows that takt losses are significantly greater for a takt time above the maximum station time (left side of Fig. 2.16) than for a takt time oriented to the weighted station workload R_n. In our example (right side of Fig. 2.18), the takt losses are close to zero for product P1, but they are still present for P2. These takt losses can be further minimized by using variable takt, in Sect. 4.5 we apply variable takt in the current example.

Critical discussion (part 1)—fluctuations of the performance level As described in Sect. 2.6.3, a worker can change his/her performance level. If an uncompensated underload of a worker should actually lead to a waiting time for the next order, this is usually not visible. This is because the worker recognizes this condition and counteracts it independently by reducing the performance level—he/she thus remains in a visible workflow, with unfortunately not visibly reduced performance level. Waiting times are often perceived as "unpleasant" by the workers because they

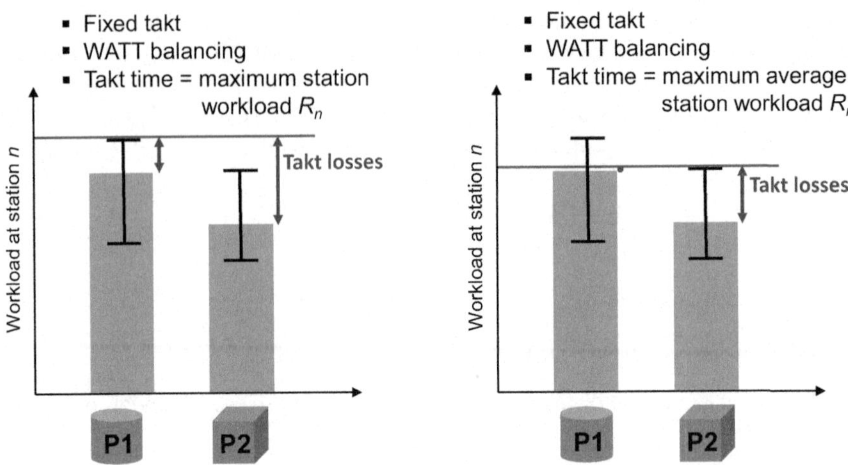

Fig. 2.16 Takt losses for two products in mixed-model assembly of type 3

are visible. The worker has no influence on this waiting time, it is immanent in the system due to balancing, but cannot be perceived as such by outsiders (colleagues, superiors, visitors, etc.).

To further reduce the theoretical takt losses from Fig. 2.16, both products P1 and P2 could be plotted in a T-bar diagram. The average workload of this hybrid is then lower than that of P1, the takt time could be further reduced, and the theoretical takt losses would be zero. The prerequisite is that the maximum overtakting is not limited. The takt time spread (compare Fig. 2.12) of the hybrid would be the maximum spread of both products. The takt time would thus be below the average load of product P1, the worker would thus basically drift at P1. The risk of model-mix losses increases significantly, and compensation possibilities, such as open station boundaries, would have to be designed to be considerably larger. In the case of this hybrid T-bar, takt losses would be shifted into model-mix losses.

2.8.2 Model-Mix Losses

Model-mix losses are losses caused by fluctuations in the workload at a given time and triggered by the prevailing order sequence. These losses are caused by the use of the WATT method with a takt time close to the weighted station time R_n. In simple terms: By using the WATT method with a takt time smaller than the maximum station time, overload and underload situations of workers will occur. Measures must be taken in the assembly system to compensate for these overloads and underloads in order to avoid or at least minimize model-mix losses. If the maximum takt is used, model-mix losses do not occur at any time (left side of Fig. 2.16). Koehter (1986) already stated that lower model-mix losses can be "bought" with increased takt time compensation, and thus takt losses. However, he rejects this idea, since an optimization of the overall system should be aimed at.

Probably the most important method, if not a basic requirement when using the WATT method combined with takt times smaller than the maximum station time, is the use of open station boundaries (Thomopoulos, 1967; Dar-El, 1978; Koehter, 1986) as we already described in Sect. 2.6.10. Open station boundaries allow the worker to "swim" with and against the flow direction of the assembly line in a defined area. Colloquially, he/she can drift (in the flow direction) but also work ahead (against the flow direction). The effectiveness of open station boundaries has been demonstrated in numerous publications, some examples are Dar-El (1978), Boysen (2005), Koehter (1986), and Swist (2014).

The open station boundaries create a kind of buffer for the worker. If the current product in his/her station has slightly less station time than the takt time, he/she can start with the next product before the beginning of his/her station. If the following product has more station time than takt time, then he/she can use up this buffer and additionally drift into the following station. The temporary overload or underload thus does not immediately become a model-mix loss, but can be compensated—it is mastered. However, the extension of the open stations must be limited (Sect. 2.6.9), so after the "normal" station limit, the final station limit follows (see Sect. 2.6.10 and Fig. 2.14).

This game of "tides" is one of the most researched areas in mixed-model assembly (Fig. 2.17)—the assembly line sequencing problem (Thomopoulos, 1967). The method of balancing based on the experiential knowledge of planners

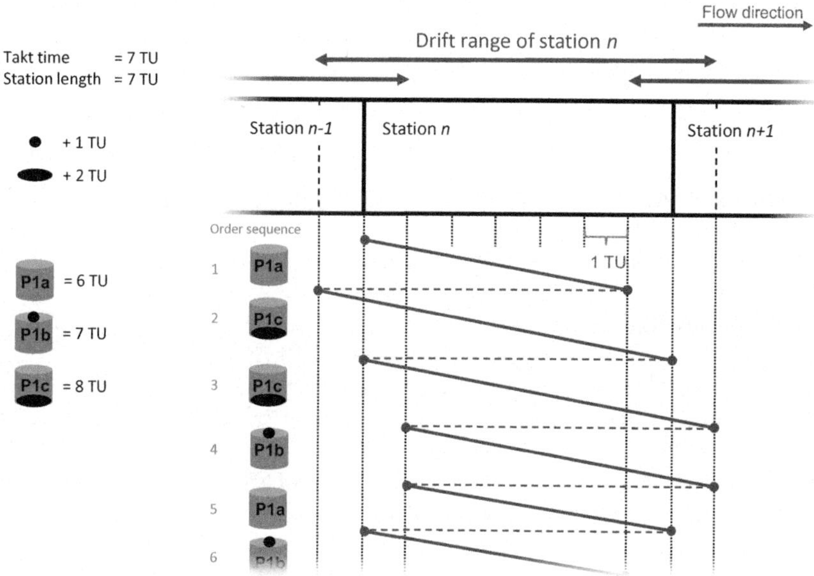

Fig. 2.17 Simulation of mixed-model assembly of type 2—simplified illustration of drift range areas with open station boundaries in fixed takt

and team leaders described in Sect. 2.6.8 plays to its strengths here, because potential drift can be taken into account during balancing. For example, WOs at successive stations are combined so that overlapping work areas do not interfere with each other as much as possible. This is information that is not included even in complete precedence graphs. With current software solutions, if the data is complete, sequence orders can be created in which the summed drift of all stations is minimized—a load-oriented series-sequence planning (Sect. 3.2.3) is created. An example of such an application is presented in Sect. 8.6.

Notes on Fig. 2.17*:*

- *In our highly simplified example, the boundary of the drift ranges is chosen to provide a buffer of +1/−1 TU.*
- *In the case of an order P1a with a station time less than the takt time, the buffer is built up. In our example, by 1 TU.*
- *This means that jobs P1c with a station time greater than the takt time can also be processed without this overload leading to a model-mix loss.*

Critical Discussion (Part 2)—Fluctuations in the Degree of Performance In all, at first purely theoretical, for the balancing of the assembly line, a constant degree of performance of the workers is assumed. In practice, however, many workers do not implement the buffers of open station boundaries in the form of a fluctuating starting point (drifting). They generally try to "work ahead" a bit (start point of work before the scheduled station start) and compensate for fluctuations in station times by adjusting their own performance level. Or in short: work is done a little faster or slower, always with the aim of not leaving the initially assigned assembly position.

The required size of these drift areas depends on several factors. Essentially these are:

- The value of the maximum permissible overtakting during balancing (Sect. 2.6.9)
- The size or cubature of the product and thus the ability whether parallel work on the product is possible
- The range of stationary equipment or machinery

Visible model-mix losses are due to:

- Waiting time of a worker in case of non-compensated underload
- Waiting times of all workers of the line, triggered by a line stop in case of non-compensated overload of a worker
- Waiting times of flexible workers (the working time of a flexible worker during an operation in the assembly line is considered productive, he/she performs work with added value to the product)
- Overtime for reassembly of work operations that could not be carried out in the assembly line due to temporary overload

In Fig. 2.18, we show an example of how model-mix losses can occur when assembling a product with two option variants in mixed-model assembly of type 2.

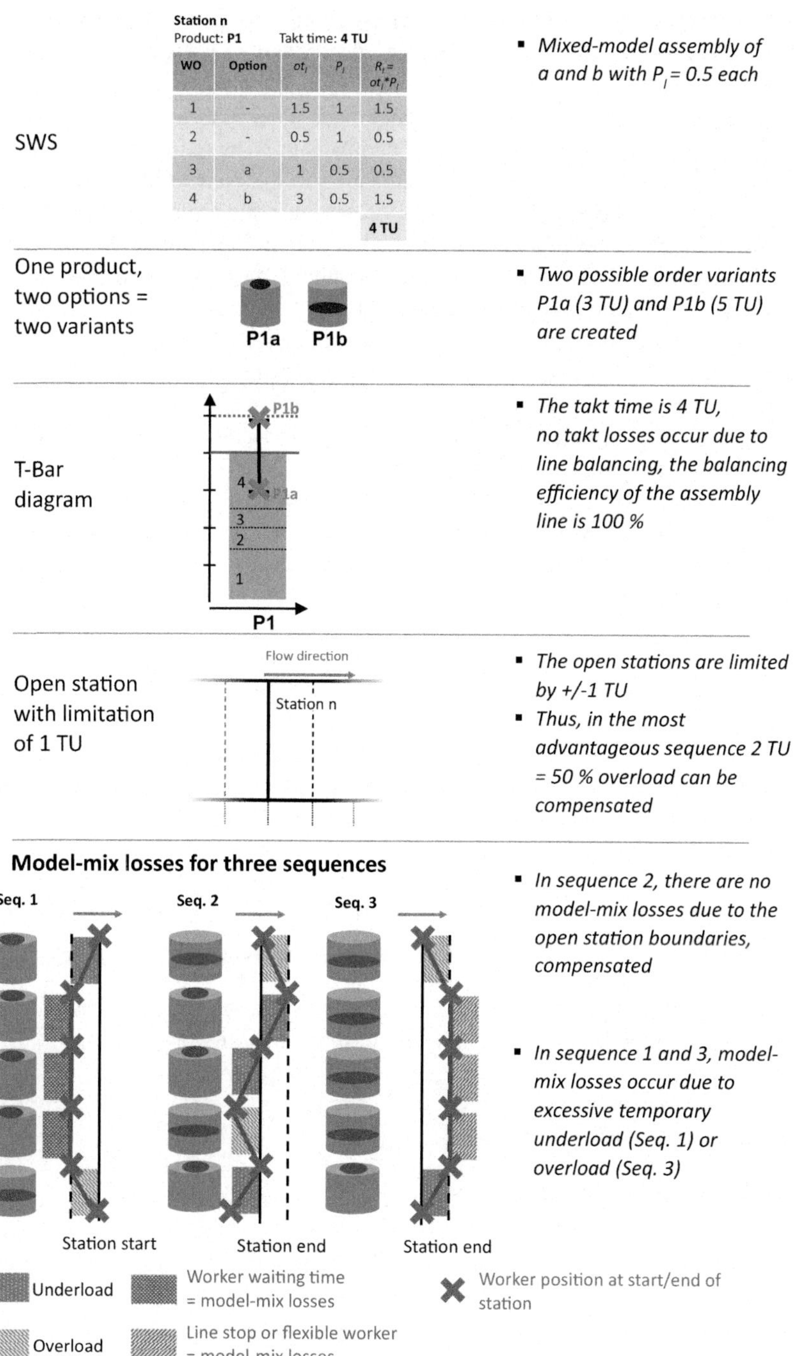

Fig. 2.18 Mixed-model assembly of type 2—occurrence of model-mix losses in the fixed takt

2.9 An Interim Conclusion on the Prevention of Utilization Losses in the Mixed-Model Assembly Design

At the end of this chapter, we should have made clear, on the one hand, the ever-increasing need to handle and master variance in a mixed-model assembly, in particular of type 3. Increasing individualization in the context of mass customization, a broader technology diversification of products and a global distribution, and thus decentralization of production, force mixed-model assemblies to master more and more variance. On the other hand, the fundamentals of flow assembly and its line balancing are explained for practical applications.

Despite decades of research and practical application of these results, utilization losses continue to be a determining cost factor in variant-rich flow assembly. Due to the increasing variance of the products to be assembled within an assembly line, existing concepts are reaching their limits. It is necessary to develop new solution concepts or to better combine existing ones.

We summarize our approaches to mastering variance in flow assemblies with our approach, the Mixed-Model Assembly Design, which we first introduced in Sect. 1.4. The strongest preventive effect is generated in the Product Design for Mixed-Model Assembly (PD for MMA, Chap. 6), utilization losses are fundamentally reduced by the design of the products, the solution space for the AD for MMA and SD for MMA is increased. In addition to the overview of MMAD from Sect. 1.4, we use the terms utilization losses, takt losses, and model-mix losses introduced in this chapter to label the Y-axes (Fig. 2.19).

The methods and concepts presented in Chap. 2 are part of assembly planning and thus part of Assembly Design for Mixed-Model Assembly (AD for MMA), which we will further strengthen with the variable takt and the VarioTakt from Chap. 4 in order to reduce utilization losses. AD for MMA is our main focus.

The focus on reducing model-mix losses is in Sequence Design for Mixed-Model Assembly (SD for MMA), which we will come to later in this book and in the following Chap. 3.

However, let us now refine our approach to AD for MMA. In Sect. 2.4.2, we described the basic four ways of dealing with option variants. If we transfer these to the context of mixed-model assemblies, four directions of work emerge:

1. Avoid or prevent options
2. Shift options outside production
3. Relocate options outside the assembly line
4. Master options, and therefore variance, in mixed-model assembly: *our focus*

A good overview of possible factors influencing utilization losses is provided by Swist (2014), which builds heavily on the work of Koehter (1986). Combined and extended with our findings, the following overview emerges (Table 2.3).

Design-for-Takt in PD for MMA does not focus on variance reduction or a general assembly time reduction, these are the domains of Variant Management and Design-for-Assembly (Table 2.3). The concepts of Design-for-Takt are intended

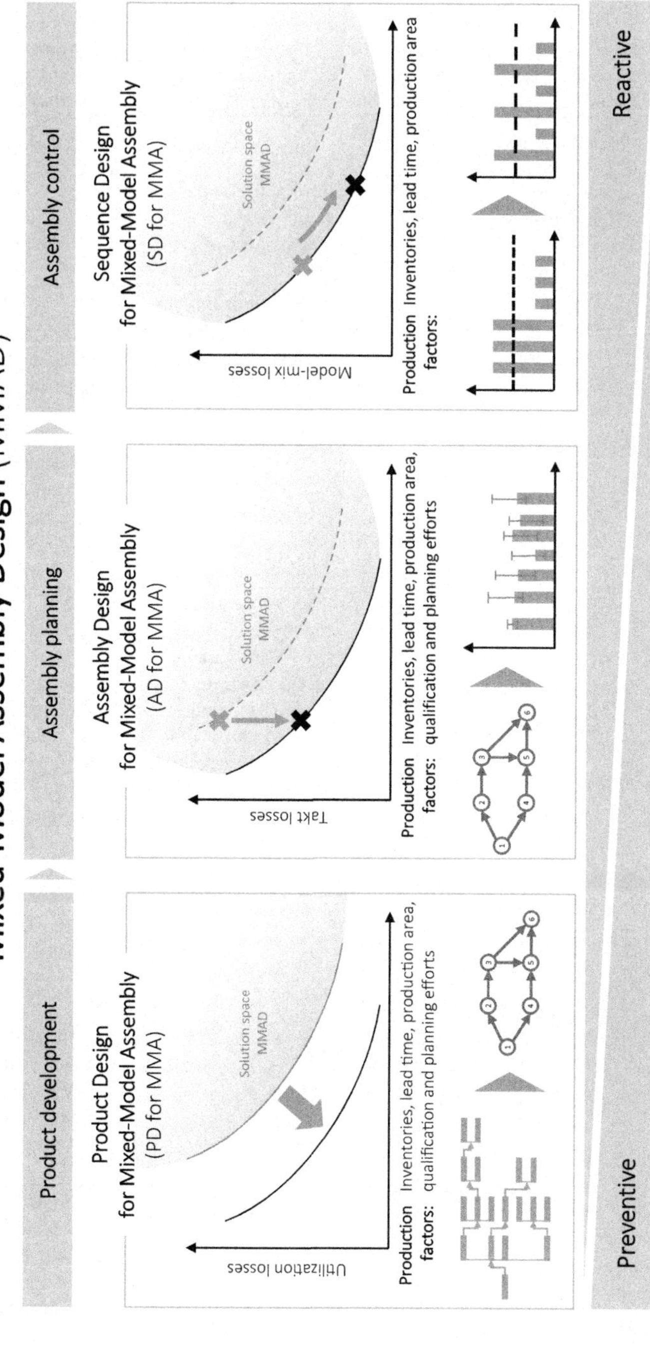

Fig. 2.19 Mixed-Model Assembly Design (MMAD) adapted from Swist (2014) and Kesselring (2021)

Table 2.3 Elements of Mixed-Model Assembly Design (MMAD) to reduce workload losses

<table>
<tr><th>Target: Reduction of utilization losses</th><th>Avoid/prevent variance</th><th>Relocate variance outside production</th><th>Relocate variance outside the assembly line</th><th>Master variance in the mixed-model assembly</th></tr>
<tr><td rowspan="4">PD for MMA (Chap. 6)</td><td colspan="3">• Design-for-Assembly (DFA)
• Variant Management</td><td rowspan="2"></td></tr>
<tr><td>– Standardization norm and common parts</td><td colspan="2">– Modularization, modular design, and platform principle
– Variance through software</td></tr>
<tr><td colspan="2" rowspan="2"></td><td colspan="2">• Design-for-Takt (Sect. 6.3)</td></tr>
<tr><td>– Shift variance to pre-assembled components</td><td>– Flexibilization of precedence relationships
– Small operation times, uniformly distributed, levelled
– Workload distribution assembly assignment space</td></tr>
<tr><td>AD for MMA-Structure (Chap. 2)</td><td colspan="2"></td><td>• Vertical segmentation
– Shift variance to takt time decoupled pre-assemblies
– Modular and hybrid structures</td><td>• Horizontal segmentation
– Parallelization of assembly lines, stations, activities
– Modular and hybrid structures</td></tr>
<tr><td>AD for MMA-Work organization (Chap. 2)</td><td colspan="3"></td><td>– Open stations
– Cycle modules (Sect. 8.3)
– Teamwork
– Flexible workers
– Special task workers (Sect. 7.1.3)
– . . .</td></tr>
<tr><td>AD for MMA-Balancing (Chaps. 2, 4, 5)</td><td colspan="3"></td><td>– WATT balancing (Sect. 2.6)
– Variable takt (Sect. 4.4)
– VarioTakt (Sect. 7.1)</td></tr>
<tr><td>SD for MMA—sequence planning (Chap. 3)</td><td colspan="3"></td><td>• Heijunka
– Level scheduling
– Car sequencing
– Mixed-model sequencing</td></tr>
</table>

to manage the remaining variance in assembly and to increase the solution space for the approaches of AD for MMA (Fig. 2.19). In Chap. 6, we provide a brief overview of possible Design-for-Takt concepts.

The assembly control methods in the SD for MMA (at the bottom of Table 2.3) are to be considered reactive; nevertheless, they form an important possibility to reduce model-mix losses. Takt losses can no longer be reduced at this stage. Sequencing can improve the overall assembly system conditions, but this is at the expense of other metrics such as lead time or inventories. Research results from the field of sequencing are currently only being applied in a few, mostly larger companies, but with an increasing tendency (Sect. 8.5). From our own experience and feedback from many other medium-sized and smaller assembly plants, we can still assume a manual sequence planning process for the most part, mostly on the basis of car sequencing, as we describe in Sect. 3.2.3.

We will focus on managing variance in mixed-model assembly using methods from assembly planning, AD for MMA. Thus, we are on a medium- and long-term planning horizon with preventive as well as reactive measures. Innovations in the area of assembly structures are currently given a lot of importance. For example, modular or hybrid assembly layouts are currently a focus of scientific research and corporate initiatives, examples of which are shown in Chaps. 7 and 8. Further variances can be compensated for with methods from the AD for MMA, such as work organization or segmentation.

The method of WATT balancing described in this chapter, but also variable takt and VarioTakt, are balancing methods from AD for MMA. We deliberately focus on the concept of variable takt, since it has received almost no attention in the scientific literature or in practical applications. From now more than 5 years of practical application in a large automotive-related assembly plant with 10 different products and an annual volume of 20,000 units, we are convinced of the benefits and practicality of variable takt. We will show that variable takt can not only significantly reduce takt losses but also *make it possible* to produce very different products (in terms of structure and assembly effort) on a clocked assembly line.

In the end, it is always about balance! As we described in Sect. 2.8.2, all measures ultimately aim to continuously load the worker on the assembly line with the same intensity. This generates the lowest utilization losses and is the best prerequisite for good quality in the assembly process—a realization shared by all experienced assembly experts. For this reason, in the next chapter we will look at the basics of "levelled loading" or "levelling" of an assembly line—Heijunka.

References

Altemeier, S. (2009). *Cost-optimal capacity coordination in a clocked variant flow line* (227 p.). University Library of Paderborn. http://digital.ub.uni-paderborn.de/ubpb/urn/urn:nbn:de:hbz:466-20090608028

Baybars, I. (1986). A survey of exact algorithms for the simple assembly line balancing problem. *Management Science, 32*(8), 909–932. https://doi.org/10.1287/mnsc.32.8.909

Boysen, N. (2005). *Variantenfließfertigung* (294 p.). Springer. https://www.springer.com/gp/book/9783835000582

Boysen, N., Fliedner, M., & Scholl, A. (2007). A classification of assembly line balancing problems. *European Journal of Operational Research, 183*(2), 674–693. https://doi.org/10.1016/j.ejor.2006.10.010

Boysen, N., Fliedner, M., & Scholl, A. (2009a). Assembly line balancing: Joint precedence graphs under high product variety. *IIE Transactions, 41*(3), 183–193. https://doi.org/10.1080/07408170801965082

Boysen, N., Fliedner, M., & Scholl, A. (2009b). Sequencing mixed-model assembly lines: Survey, classification and model critique. *European Journal of Operational Research, 192*(2), 349–373. https://doi.org/10.1016/j.ejor.2007.09.013

Cachon, G., & Terwiesch, C. (2020). *Operations management* (768 p.). McGraw Hill.

Da Silveira, G., Borenstein, D., Fogliatto, S., & F. (2001). Mass customization: Literature review and research directions. *International Journal of Production Economics, 72*, 13. https://doi.org/10.1016/S0925-5273(00)00079-7

Dar-El, E. M. (1978). Mixed-model assembly line sequencing problems. *Omega, 6*(4), 313–323. https://doi.org/10.1016/0305-0483(78)90004-X

Decker, M. (1993): *Variantenfließfertigung* (229 p.). Schriften zur quantitativen Betriebswirtschaftslehre Physica-Verlag.

Deming, W. E. (1982). *Out of the crisis* (524 p.). MIT.

Domschke, W., Scholl, A., & Voß, S. (1997). *Produktionsplanung: Ablauforganisatorische Aspekte* (455 p.). Springer.

Fattahi, P., & Salehi, M. (2009). Sequencing the mixed model assembly line to minimize the total utility and idle costs with variable launching interval. *The International Journal of Advanced Manufacturing Technology, 45*. https://doi.org/10.1007/s00170-009-2020-0

Fujimoto, T. (1999). *Evolution of manufacturing systems at Toyota* (400 p.). Taylor & Francis.

Gans, J. E., Lanza, G., Müller, R., Peters, S., & Schoen, L. (2011). Prognose des Driftverhaltens getakteter Montagelinien. *VDI Fachmedien*, 162–166.

Günther, O., & Tempelmeier, H. (2012). *Produktion und Logistik* (388 p.). Springer.

Halubek, P. (2012). *Simulation-based planning support for variant flow manufacturing* (191 p.). Vulkan Verlag.

Jackson, J. R. (1956). A computing procedure for a line balancing problem. *Management Science, 2*(3), 197–286. https://doi.org/10.1287/mnsc.2.3.261

Kesselring, M. (2021b). Product design for mixed-model assembly lines. Master Thesis, WHU, Otto Beisheim School of Management.

Klindworth, H., Otto, C., & Scholl, A. (2012). On a learning precedence graph concept for the automotive industry. *European Journal of Operational Research, 217*(2), 259–269. https://doi.org/10.1016/j.ejor.2011.09.024

Koehter, R. (1986). *Verfahren zur Verringerung von Modell-Mix-Verlusten in Fließmontagen* (182 p.). Springer.

Koren, Y. (2010). *The global manufacturing revolution: Product-process-business integration and reconfigurable systems* (399 p.). Wiley.

Koren, Y. (2021). The local factory of the future for producing individualized products. *The BRIDGE, 51*(1), 100 p. https://www.nae.edu/251191/The-Local-Factory-of-the-Future-for-Producing-Individualized-Products

Kotha, S. (1995). Mass customization: Implementing the emerging paradigm for competitive advantage. *Strategic Management Journal, 16*, 21–42. https://doi.org/10.1002/smj.4250160916

Kratzsch, S. (2000). *Prozess- und Arbeitsorganisation in Fließmontagesystemen* (237 p.). University of Braunschweig.

Liker, J. K. (2004). *The Toyota Way* (330 p.). McGraw Hill

Liker, J. K. (2006). *Der Toyota Weg* (438 p.). Finanzbuch Verlag.

Liker, J. K. (2021). *The Toyota Way* (449 p., 2nd ed.). McGraw Hill.

Loch, C., Sting, F., Bauer, N., & Mauermann, H. (2010). The globe: How BMW is defusing the demographic time bomb. *Harvard Business Review Online,* March 2010. Accessed June 03, 2021, from https://hbr.org/2010/03/the-globe-how-bmw-is-defusing-the-demographic-time-bomb

Moench, T., Huchzermeier, A., & Bebersdorf, P. (2020a). Variable Takt times in mixed-model assembly line balancing with random customization. *International Journal of Production Research.* https://doi.org/10.1080/00207543.2020.1769874

Moench, T., Huchzermeier, A., & Bebersdorf, P. (2020b). Variable takt time groups and workload equilibrium. *International Journal of Production Research.* https://doi.org/10.1080/00207543.2020.1864836

Ohno, T. (2013). *The Toyota production system* (176 p.). Campus Verlag.

Otto, C., & Otto, A. (2014). Multiple-source learning precedence graph concept for the automotive industry. *European Journal of Operational Research, 234*(1), 253–265. https://doi.org/10.1016/j.ejor.2013.09.034

Pine, B. J., Bart, V., & Boynton, A. C. (1993). Making mass customization work. *Harvard Business Review Online.* Accessed April 29, 2021, from https://hbr.org/1993/09/making-mass-customization-work

Pröpster, M. H. (2015). Methodik zur kurzfristigen Austaktung variantenreicher Montagelinien am Beispiel des Nutzfahrzeugbaus (238 p.). utzverlag.

Rother, M. (2009). *Toyota Kata* (400 p.). McGraw-Hill.

Rother, M. (2013). *Die Kata des Weltmarktführers: Toyotas Erfolgsmethoden* (299 p.). Campus Verlag.

Salveson, M. E. (1955). The assembly line balancing problem. *The Journal of Industrial Engineering, 6*, 18–25.

Schlick, C., Bruder, R., & Luczak, H. (2010). *Arbeitswissenschaft* (1194 p.). Springer.

Scholl, A. (1999). *Balancing and sequencing of assembly lines* (334 p.). Physica-Verlag.

Shtub, A., & Dar-El, E. M. (1987). A methodology for the selection of assembly systems. *Int J Prod Res, 27*(1), 175–186. https://doi.org/10.1080/00207548908942537

Spur, G., & Stöferle, T. (1986). *Handbuch der Fertigungstechnik* (820 p.). Carl Hanser Verlag.

Sternatz, J. (2010). Automatic clocking without creating a precedence graph. *MTM User Conference*, 04/16/2010.

Sugimori, Y., Kusunoki, K., Cho, F., & Uchikawa, S. (1977). Toyota production system and Kanban system materialization of just-in-time and respect-for human system. *The International Journal of Production Research, 15*(6), 553–564. https://doi.org/10.1080/00207547708943149

Swist, M. (2014). *Taktverlustprävention in der integrierten Produkt- und Prozessplanung* (328 p.). Apprimus Verlag.

Thomopoulos, N. T. (1967). Line balancing-sequencing for mixed-model assembly. *Management Science, 14*(2), B-59-B-75. https://doi.org/10.1287/mnsc.14.2.B59

Weiss, C. (2000). *Methodengestützte Planung und Analyse von Endmontagelinien in der Automobilindustrie* (226 p.). University of Karlsruhe.

Wiendahl, H.-P., Gerst, D., & Keunecke, L. (2004b). Variantenbeherrschung in der Montage. *Springer Verlag.* https://doi.org/10.1007/978-3-642-18947-0

Wildemann, H. (2020). Variantenmanagement: Leitfaden zur Komplexitätsreduzierung, –beherrschung und -vermeidung in Produkt und Prozess. Transfer Center for Production Logistics and Technology Management.

Witte, H. (2007). *Allgemeine Betriebswirtschaftslehre* (319 p.). Oldenburg Verlag.

Womack, J. P., Jones, D. T., & Roos, D. (1990). *The machine that changed the world* (323 p.). Westview Press.

3 Heijunka: Fast like a Tortoise

In the third chapter, we will continue to approach the core topic of variable takt and VarioTakt. This section is once again intended as a basic chapter to generate a broad and deep understanding of the balancing of flow assemblies. It focuses on elements of Sequence Design for Mixed-Model Assembly (SD for MMA) from our MMAD model. As before, in this part, we will try to combine the approaches of the scientific literature on levelling with management-oriented practical knowledge on Heijunka to generate added value. We will show that the origins of Heijunka can be found in Toyota's Lean Management philosophy and thus form the basis for almost every production system used today. In parallel, levelling has been and continues to be a focus of much scientific work in the field of Operations Research. It will become apparent that variable takt complements existing concepts for levelling production quite decisively and creates a *self-levelling* production system. For example, VarioTakt takes on the task of regulating overload situations in an assembly line without intervening in the series-sequence. As a result, series-sequence planning could be focused exclusively on supply levelling—or true build-to-order production could be implemented (Fig. 3.1).

3.1 Historical Development from the Toyota Production System

"We would rather be like a tortoise, slowly and steadily reaching our goal than hastily and quickly like the hare" (Ohno, 2013, p. 102). For several decades this guiding principle has characterized the production community oriented according to Lean principles; variable takt helps to put this guiding principle into practice. What at first seems contradictory acquires its profound meaning upon understanding Taiichi Ohno's world of thought and forms one of the most significant foundations for the implementation of Lean Management (Fujimoto, 1999; Ohno, 2013), which finds its roots in the Toyota Production System (TPS). For this reason, we would first

P. Bebersdorf, A. Huchzermeier, *Variable Takt Principle*, Management for Professionals, https://doi.org/10.1007/978-3-030-87170-3_3

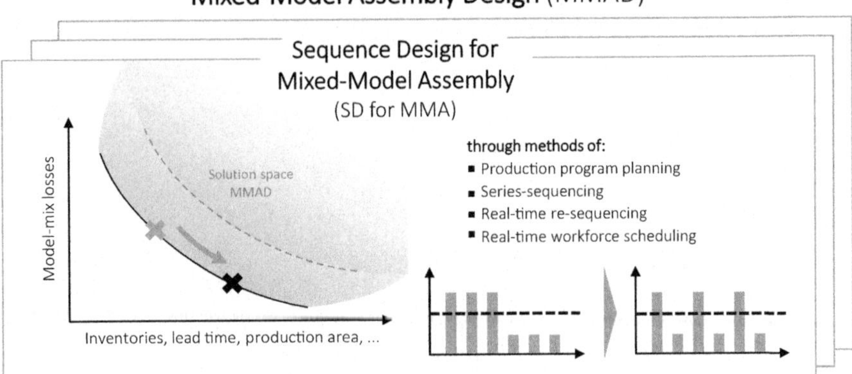

Fig. 3.1 Sequence design for mixed-model assembly in MMAD

like to explain the basic structure of the TPS, but without going further into detail about individual methods or even tools.

Many Lean Management experts may regard the following descriptions as "old hat," but even after many years of applying this philosophy (we use this term here on purpose), the basic idea behind it does not lose its meaning and impact. When management gets caught up in the day-to-day problems of production and can no longer see the forest for the trees, it often helps to take a few steps back, focus on the essential contents of the TPS and thus reassess the situation.

At this point, it should be pointed out that, when viewed superficially, the TPS, as well as the Lean Management approach itself, are often seen as a collection of tools and methods. Copying individual elements very often fizzles out ineffectively and is then seen as proof of non-applicability in one's own organization. Here we try to emphasize its holistic meaning, its interdependencies, which only unfold their full effect in interaction. Thus, the TPS can be understood both "as a method, a process or program, a strategy, a goal, a state of belief, or a philosophy" (Lee & Jo, 2007, p. 3667). Thus, it is more important to understand the TPS in its entirety, to adapt it accordingly, and not to copy it—and certainly not just some of its elements. For this reason, the Toyota Production System is very often depicted in the form of a house—all elements are needed for a complete house. In the meantime, there are numerous visualizations of the TPS; in the following, we use a very simplified form in order to focus on the essential core elements.

The roof of the house reflects the goals of the production system. Not surprisingly, this is the magic triangle of quality, cost, and delivery lead time. Toyota adds to these the goal of safety for all employees and the pursuit of high employee motivation. In the meantime, almost all production systems have adopted these goals for themselves. The TPS is based on the assumption that these goals will be achieved if all waste is eliminated from all processes. Waste is defined as any activity that does not add value to the final product. The TPS distinguishes the following seven types of waste (Ohno, 2013):

- Overproduction—the mother of all waste
- Waiting time
- Transport
- Unnecessary/inappropriate processes
- Inventories
- Unnecessary movements
- Defects

There are countless practical examples of how reducing these wastes leads to improvements in quality, cost, and delivery lead time. Some are obvious, such as the fact that transporting products is a hassle and therefore increases costs, or that waiting times increase lead times and therefore have a negative impact on delivery lead time. Other connections are not immediately obvious at first, such as the risk of workplace accidents increases when a lot of material is moved or transported from one part of the factory to another. Overproduction degrades the quality of products because they have to be transported to a warehouse in addition, which increases the risk of damage. Practitioners should find it easy to cite other examples.

To avoid waste, Toyota's Production System is essentially built on two pillars: Jidoka and Just-in-Time.

The term autonomation is often used as a synonym for Jidoka. Toyota understands this to mean the endeavor to always and immediately produce the right quality. In the event of errors, which are understood as deviations from a defined standard, machines should stop automatically or assembly workers should stop the assembly line. Errors must cause pain. After all, why stop at a defect when it is possible to outfeed the production units at any point without much effort and to fix them without time pressure. Defects, or errors, should be made transparent immediately in order to eliminate their root cause in the long term. In TPS, one of the most important goals is to ensure that no defect is passed on to the subsequent process.

As shown in Fig. 3.2, standardization and visualization form the foundations of Jidoka implementation. Without standardization and its visualization, a deviation, and thus defect, is not recognizable.

The second pillar for achieving the goals of the TPS is the Just-in-Time (JIT) principle. Processes should flow continuously, always pulling on the upstream process and running in a defined takt. We have already discussed the advantages of a continuous product flow several times. The foundation for JIT is standardization and Heijunka (Ohno, 2013). Heijunka is the Japanese term for smoothing, but can also be translated as levelling. In the following sections, we present in more detail the importance of Heijunka as one of the main foundations of TPS. It will become apparent that when variable takt is used, a levelling of the production program, or more precisely, a levelling of the workload is achieved.

Central, and thus at the center of the TPS house, is the striving for continuous improvement, known in Japanese as Kaizen (Kai = change, Zen = for the better). In the sense of the TPS, this does not mean large innovative leaps in improvement, but rather continuous small steps, and preferably every day by every employee.

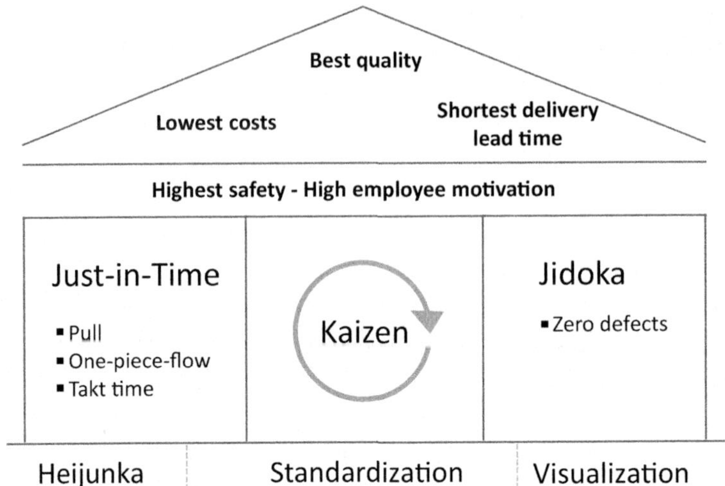

Fig. 3.2 Visualization of the Toyota Production System—adapted from Liker (2021)

This briefly explains the TPS house and thus the basis of Lean Management. In addition to standardization and visualization, Heijunka forms the foundation for the use of all the underlying Lean Management principles. Without Heijunka, these lead nowhere or cannot be implemented. Heijunka is elementary for the creation of a continuous product flow. One of its most important measures is the series-sequencing of the assembly line.

Is the TPS ultimately *only* about avoiding waste? Basically, this question can be answered with a yes. In recent decades, the understanding of how this works has deepened considerably—the goal of reducing waste always remains in all considerations (Fujimoto, 1999). While Womack et al. (1990) still described mainly individual tools that Toyota used, Liker, in what we call "the second wave of Lean understanding," recognized the principles behind the tools that generate successful implementation. Since the work of Rother (2009, 2013), the "Third Wave of Lean understanding," the Improvement and Coaching Kata has come into focus. With a continuous, targeted and scientific routine, processes, employees, and managers are continuously improved or coached based on Lean principles.

3.1.1 No Improvement Without Necessity

Common to all levels of knowledge of the TPS is that there is no improvement in the company without necessity. In the management of many production companies, this phenomenon is also well known: Thus, the greatest threat to tomorrow's success is today's success, because why change anything: "It's going on!" Already Ohno wrote in the 1990s: "In the beginning was the necessity" (Ohno, 2013, p. 48). Even at the pioneer of the application of variable takt and VarioTakt, at the tractor manufacturer

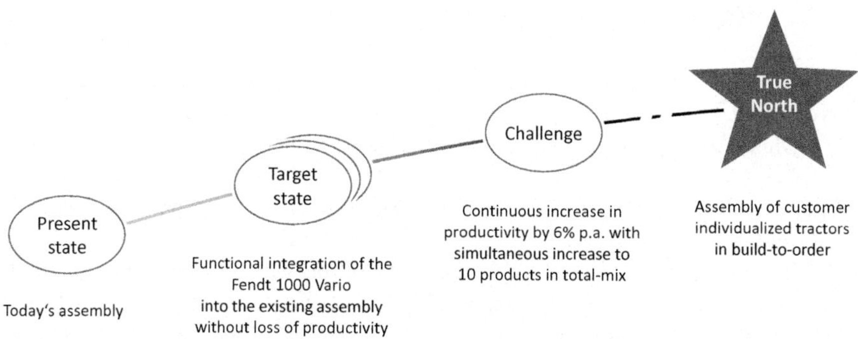

Fig. 3.3 Derivation of target states from vision, the True North, using the example of the integration of the Fendt 1000 Vario into the existing assembly structure (adapted from Rother, 2009)

Fendt, the intensity for the implementation of variable takt and the development of the VarioTakt came out of necessity. With the introduction of the Fendt Vario 1000, it was necessary to question the current takt concept. The workload of this new tractor was so great that, if a fixed takt had been maintained, the takt compensation times would have increased extremely. Assembling all tractors on one assembly line would no longer have been economically viable. However, a stand-alone assembly line for this new product was out of the question due to the small number of units (Huchzermeier et al., 2020).

Toyota recognized very early on the existence of a necessity as a prerequisite for continuous improvement. Thus, Toyota remains a very successful automotive company to this day but repeatedly puts itself in artificial crises internally to generate the need for improvement. In the Lean concepts of Improvement and Coaching Kata (Rother, 2009), the need for improvement is generated or made transparent in the first step for this reason.

Before a change to the assembly system can be started, the next target state to be achieved must be derived from the vision and a challenge. The necessity is created or presented and derived transparently for all participants (Fig. 3.3).

Conclusion The introduction of variable takt should therefore be preceded by recognition of the necessity. Without this, it will be difficult to find allies or supporters within the company. This is because large parts of the management, unfortunately, shy away from implementing changes in their own company that deviate from existing and well-known solution strategies. Variable takt has not yet found widespread use in the industrial environment, this book should contribute to change this. Advantages and experiences with the introduction of variable takt are shown in Sects. 4.5 and 4.8.

3.1.2 The Three M: Muda, Muri, and Mura

To fully understand the capabilities of variable takt to achieve true one-piece flow, it is worthwhile to look at the principle of the "3 M—Muda, Muri and Mura" from the TPS (Ohno, 2013).

Muda The Japanese term Muda has now established itself in the world of Lean Management enthusiasts as a common synonym for waste, in the sense of the seven types of waste. Many followers of the Lean philosophy see the elimination of Muda as the most important, if not the only, goal. This is because, without any waste, a process will have the lowest cost, the shortest delivery lead time, and the best quality. In this context, it is very important to be aware of the interdependencies of each type of waste. A reduction of one type of waste very often leads to the expansion of other types of waste. For example, waiting times at machines can be reduced by producing without demand, or batches can be increased in order to reduce set-up costs, i.e., resulting in overproduction with transports and inventories. When reducing one type of waste, the effect on all other types of waste must always be taken into account.

Muri Ohno (2013) describes Muri as the inappropriateness of the work process used. Muri further stands for overload or overuse in the Japanese language. This meaning has been adopted by almost all other authors and companies to explain the meaning of Muri. An overload (Muri) of people and machines can lead to waste, Muda, in the form of defects, waiting time, or accidents. We would like to emphasize here that an overload can lead to losses, not must. By using the variable takt and the VarioTakt, these overloads can be mastered and therefore do not necessarily lead to Muda.

Underload Unfortunately, the concept of the three M's does not include a designation for underload, because this also does not initially represent waste, provided it can be mastered. Very often, an underload is immediately equated with waste and thus Muda, we expressly do not agree with this definition. This basic understanding has been developed very well in the work of Moench et al. (2020) and will be further elaborated in Chap. 5.

Mura Stands for imbalance or disequilibrium, it can be seen as the span of overload (Muri) and underload. The goal of Heijunka is to reduce Mura and thus eliminate an important basis for the creation of waste, or the seven types of waste. In a flow assembly, there are different approaches to reduce imbalances or to master them in order to avoid resulting wastes. Section 2.8.2 provides an overview of these approaches.

By means of a schematic T-bar representation, based on the definitions from Chap. 2, the relationships of the three M can be well visualized (Fig. 3.4). Mura can be understood as the span of Muri and underload. If the takt time is greater than the weighted workload R_n, takt losses initially occur in the form of waiting times (see

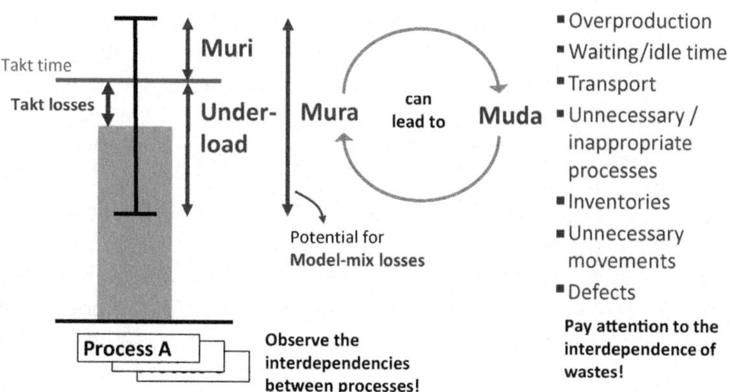

Fig. 3.4 Muri, Underload, Mura, and Muda

Sect. 2.8.1). If the underlying probabilities of occurrence of the individual option variants occur as planned, these takt losses cannot be avoided. With a deviating model-mix, the underloads of individual orders can lead to further waiting times if they are not mastered (see Sect. 2.8).

These underloads can not only show through as losses in the form of waiting times. Large underloads increase the susceptibility of a production process to defects; this effect initially appears contradictory. In practice, it has been recognized that qualitatively the best work results are achieved when employees are in a continuous workflow that neither overloads nor underloads them (Diwas & Terwiesch, 2009). If an employee has to regularly interrupt his/her workflow because there is not enough work content due to an underload situation, he/she is distracted, leaves his/her structure or rhythm—defects occur.

Again, it is very obvious that if the Mura is too large, an excessive overload of the worker, defects can occur more easily or the risk of line stoppages is increased. For a single assembly process in a flow assembly line, it can be fundamentally assumed that a reduction of Mura leads to a reduction of Muda, in the form of utilization losses (Sect. 2.8).

In a real application, it should be noted that in the vast majority of cases a production process consists of several individual processes running serially and/or in parallel. Due to the interdependence of the different types of waste in different production stages, a reduction of Mura or Muda in process A can lead to a deterioration in one of the other processes (Fig. 3.4). The total waste could thus increase, for which reason entire value streams should be considered and improved as a matter of principle. Heijunka should be applied to the entire manufacturing process.

And If You Could Choose: Would You Rather Have Takt Losses or Model-Mix Losses?

This question cannot be answered in a generalized way; it must be decided in the context of the respective production environment and, supported by optimization algorithms in balancing and series-sequencing. It remains a management decision. Takt losses occur in the balancing phase and take effect over the entire runtime of the assembly system in the form of waiting times. If, during the operation of the assembly system, the probabilities of occurrence of option variants occur as planned in the balancing phase, then the takt loss becomes effective at all times. Model-mix losses occur during the operation of the assembly system when overload and underload can no longer be compensated due to open station boundaries and drifts (see Sect. 2.6.10). They can occur, but do not have to.

For example, one strategy might be to minimize takt losses from the beginning and set the takt as close as possible to the weighted average of station times, thus no loss would initially be planned into the assembly system. With this step, the overload, Muri, is taken to the extreme and the risk for model-mix losses increases—but Mura remains the same. With series-sequencing strongly oriented to overload (Sect. 3.2.3), model-mix losses could be kept small. With this approach, however, it should be noted that takt losses only ever take effect in the waiting time of *one* worker; overload, due to insufficiently levelled series-sequencing, leads to a line stop and thus to waiting times for *all* workers in the assembly line or to the need to use flexible workers.

3.2 Heijunka

In order to understand why Heijunka is the basis for the Just-in-Time pillar of the TPS, we have to take a brief thinking step back to the 1950s of the Toyota company. Toyota had recognized that in its situation it was necessary to produce in flow and also only the (pre-)material that was consumed. Only in this way was it possible to produce a large number of different products in small volumes with limited resources. Inspired by American supermarkets, the pull principle was thus transferred to production (Ohno, 2013). One of the biggest problems, in the implementation of the pull principle by supermarkets in production, arose from fluctuating withdrawal quantities in supermarkets. This effect is known to many in theory and practice as the Bullwhip effect (Lee et al., 1997, Chen et al., 2000) small fluctuations at the end of a supply chain create ever-increasing fluctuations against the direction of flow. These fluctuations significantly disrupted the upstream processes that produced the sampled parts, creating the need for production levelling. The basic method in TPS to implement this levelling by pull in the whole production network is Kanban. However, a detailed description is not part of this book, here we refer to the classic literature by Womack et al. (1990) and Liker (2021).

On closer inspection, Heijunka is one of the most contradictory elements (Womack et al., 1990; Liker, 2021) in a production system; it is initially in clear conflict with true build-to-order production. This is because, in most implementations of Heijunka, inventories and buffers are used to decouple the production sequence from true demand. For this, it is a concept that represents the best compromise between minimizing manufacturing costs and that of a short delivery lead time. The output of produced goods is thus deliberately decoupled from the actual demand. This initially creates the impression that it must be accepted that some customers will have to wait longer for their product in order to improve the levelling of production (Liker, 2004). However, this impression is deceptive because a complete focus on a build-to-order sequence in today's traditional production organizations not only generates higher manufacturing costs but does not necessarily have a shorter delivery lead time with current production methods. In order to be able to manufacture any product at any time in any quantity, very elaborate arrangements must be made. Inventories of input materials for all products must be kept at a maximum, and machine and personnel capacities would always have to be designed for the maximum. Employees must be available to work in a wide variety of production areas at short notice. It is easy to understand that such a system very quickly reaches its limits, causes high costs, produces poorer quality, and ultimately, due to its susceptibility to faults, realizes longer delivery lead times.

But there is hope! It can be seen that current developments in the use of artificial intelligence to forecast manufacturing flows, the digitization of factories, the expansions of additive manufacturing and the further development of assembly concepts such as variable takt and VarioTakt will lead to being able to develop much more in the direction of a true one-piece-flow with a true build-to-order sequence. At the same time, variable takt alone does not untie the Gordian knot between levelled loading the system and directly satisfying customer demand. **However, the VarioTakt can contribute significantly to implementing a levelled load on the assembly line and a true build-to-order sequence at the same time**. Already in Sect. 2.1, we described the inability of the fixed takt to produce an even flow of workload for all workers for products with varying total assembly workload. In this chapter and in Chap. 4, we will demonstrate how a levelled workload, i.e., a balanced and continuous flow of workload, can be produced when variable takt is used despite different products. In addition to the classic approaches from the TPS to create Heijunka by a targeted modification (i.e., use of repeated small batches or sequence patterns) of the assembly sequence, variable takt is another concept to implement Heijunka in a highly customer-centric assembly with a wide variety of products.

3.2.1 Benefits of Levelled Production

Levelled production, not only in the assembly area, offers numerous advantages for the company and its customers. It should be pointed out again at this point that Heijunka or a levelled workload is an advantage in itself—everything is literally *in flow*, we have described the advantages in Sect. 2.1. However, this advantage is often bought with waste and this to the detriment of true build-to-order production.

Benefits of levelled production are:

Prerequisite for Lean Management

Heijunka creates the most important basis or prerequisite of the vast majority of Lean methods (Liker, 2021) and supports the application of the Lean philosophy in total.

Without levelled production, the application of other Lean methods is not purposeful or possible (Ohno, 2013). This characteristic probably describes the most important overarching benefit of levelled production.

Inventory Reduction
By using Heijunka, and specifically Kanban, inventories can be reduced throughout the manufacturing process (Veit, 2010). Inventories of finished goods can be kept constant to meet unexpected customer demands. In return, with the introduction of Heijunka and Kanban, inventories in the manufacturing process are reduced (Liker, 2021).

Reduction of the Bullwhip Effect
A levelled last production step (usually assembly) significantly reduces the Bullwhip effect (Veit, 2010). This is done by variance of the ordered parts, i.e., batching of orders no longer takes place (Lee et al., 1997).

Minimizing the Risk of Large Finished Goods Inventories (Liker, 2021)
If Heijunka is dispensed with and production is organized in large batches, there is a considerable risk of excessive capital and inventory tie-up or write-offs if the sales volume is incorrectly forecast.

Short Delivery Lead Times = Short Order-to-Cash Cycles
If each product is manufactured regularly and not only at specific time intervals, short-term individual customer demands can also be met. This significantly improves the company's cash flow and reduces the need for working capital.

Improved Quality Through Short Delivery Lead Times
Heijunka can reduce inventories of finished goods, but also of intermediate products (Veit, 2010). This has a positive effect on the quality of the products. Inventories have to be stored and transported. Damage can occur in all these unnecessary process steps; by reducing these process steps, quality reductions can be avoided.

Reduction of Faulty Production
Since levelling eliminates large batches, defects or defective pre-products are detected more quickly. A possibly ongoing faulty production process is detected earlier and thus generates less scrap.

Improved Assembly Line Productivity
By ensuring an even load on all workers in an assembly line, there will be fewer disturbances (especially in overload situations) and therefore line stoppages, which represent a loss of productivity.

With a levelled workload, station times can much better approximate the takt time, takt compensation is reduced, and there is less idle or waiting time.

Improved Productivity Through Higher Equipment Availability
Levelled production can reduce load peaks on machines and equipment. Load peaks increase wear on machinery and equipment and increase the risk of breakdowns.

3.2.2 Heijunka in the Sense of the Toyota Production System

In the understanding of the TPS, and thus of most production systems, Heijunka stands for a levelling of the production volume and the production mix (Liker, 2021). Orders are not processed according to the sequence of how they arrive but are collected for a period of time and then evenly distributed. Thus, production is not done in true customer demand but also not in large customer-neutral batches. If batches are produced without a customer order, then this is referred to as build-to-inventory. In classic batch production, either large quantities are pre-produced purely on a forecast or customer orders are collected until the desired batch size is reached. This is all done with the desire to achieve economies of scale in fabrication and assembly. This may still be the case in the production of customer-neutral and (pre-)products with low added value, but in the vast majority of assemblies, this is no longer the case. Mix variants in Heijunka, the production of real customer orders plus forecast orders, can be used to level out further fluctuations in demand.

Toyota recognized early on that large batch production had four major disadvantages (Liker, 2021):

1. Customers are not always predictable in their demand. Despite great efforts in market analyses and customer surveys, customer demand can fluctuate significantly, also triggered by unforeseeable external influences. There is a risk that customer demand for a particular product cannot be satisfied or may disappear altogether.
2. Inventories and inventory costs of finished goods can fluctuate extremely if customers temporarily stop taking the expected quantities.
3. Different products usually occupy different resources in terms of type and volume. Or in short: "Smaller" products usually require fewer employees for production than "larger" products. If the products are grouped together in large batches, the result is a very unbalanced load on production capacity—one day there are too many employees in the production halls, the next day too few.
4. It can be assumed that different products in an assembly require different preliminary products. These are either provided by the company's own upstream processes (e.g., fabrication) or by external suppliers. If the products are assembled in large batches, large and unevenly distributed requirements arise in the upstream processes. The upstream processes can adjust to this in two ways: (1) they keep enough capacity available to meet the temporary large requirements at short notice. In other periods, these capacities (machines and employees) are not used, which is a big waste. Or (2) they try to produce the demands evenly and store the finished or intermediate goods in a large buffer—also a type of Muda. In both scenarios, a small change in the assembly production program can lead to a

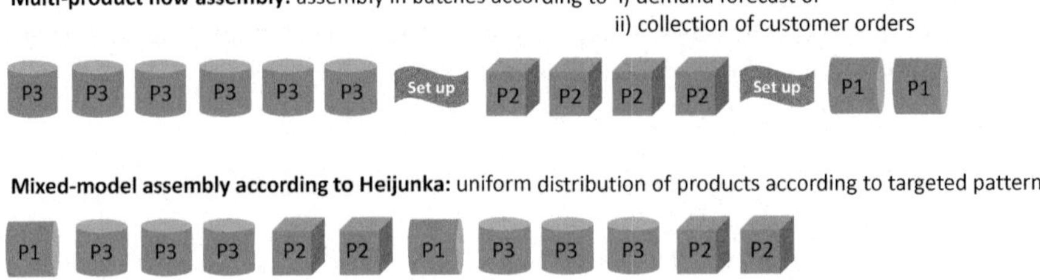

Fig. 3.5 Assembly in batches vs. mixed-model assembly using Heijunka

large upswing in the upstream process and their upstream processes (Veit, 2010). In most cases, the wrong upstream products were produced, and the supply chain to assembly breaks down.

Put simply, Heijunka can be understood in terms of TPS as adaptive production in small batches with targeted patterns. The production of large quantities of products on pure forecast information is avoided by summarizing real customer requirements for short periods (several hours to one day), dividing them into small batches and distributing them evenly in the production program. In order to promote a levelling of the workload in the assembly line, this is done according to a targeted pattern; for example, after two units of the complex product P1, four units of product P3 (product with a low workload) must always be scheduled. If the order quantities of the individual products fluctuate or change, the Heijunka batches can be adjusted in the targeted pattern (Fig. 3.5).

A prerequisite for the implementation of Heijunka in assembly is the minimization, or better the elimination, of all setup processes. For this reason, methods for minimizing setup times (e.g., SMED = Single Minute Exchange of Dies) developed as one of the core capabilities for implementing Heijunka in the TPS.

Conclusion With Heijunka, in the sense of the TPS, the focus is on an evenly distributed volume per unit of time through uniformly distributed products (cf. Sect. 2.4.1). Option variants are initially not in the consideration set of Heijunka according to TPS. If all option variants were included in the application of Heijunka, the application would become extremely complex and thus no longer applicable. For this reason, Toyota tried for a very long time to keep its offered option variants as small as possible, a clear disadvantage compared to the increasingly individualized product range of other, especially German, car manufacturers. With increasing mass customization (Pine et al., 1993; Kotha, 1995; Da Silveira et al., 2001), Heijunka's TPS approach is reaching its limits. In a future true personalized production, advanced concepts must be used to load production evenly.

3.2.3 Heijunka Through Series-Sequence Planning: The Classic Approach

Production program planning (Boysen, 2005) precedes any series-sequence planning and is therefore assumed in the following. In production program planning, the type and quantity of products to be produced in assembly and their variants are defined. The main task of program planning is, on the one hand, the synchronization of production quantities with customer requirements and, on the other hand, the temporal levelling of these production quantities (Swist, 2014). The result is usually given to series-sequencing in the form of shift programs (Domschke et al., 1997). The task of the latter is to bring this shift program into as optimal an assembly sequence as possible, depending on constraints and on the basis of defined target criteria (Boysen, 2005).

Kilbridge and Wester (1966) and Thomopoulos (1966, 1967) were the first authors to describe the series-sequencing problem and develop solution procedures. The literature distinguishes two main objectives of series-sequence planning (Boysen, 2005):

- *Overload-oriented approach*: In the context of line balancing, especially when using the WATT method (Sect. 2.6), overloads (Muri) and underloads occur with orders. The goal of sequencing is to minimize overruns of the final station boundaries by alternating orders with overloads and underloads, drifting, and using open station boundaries. In a multi-variant mixed-model assembly line of type 3 with 200 stations and a shift production of over 100 units, this is an extremely complex task.
- *Supply-oriented approach*: This approach is also called Just-in-Time-oriented approach (Boysen, 2005), its origins are in TPS. Different products and different option variants contain different components. These are provided by upstream processes, such as pre-assembly, fabrication or suppliers. An extreme case is Just-in-Time delivery or production. It is desirable to level out this demand for upstream products (see Sect. 3.1). For this reason, orders should be placed in the assembly line in such a way that there is a levelled demand for the different upstream products.

Due to the two different objectives of series-sequence planning, three approaches have developed in literature and practice (Boysen, 2005):

Level Scheduling

The origins of this method lie in the TPS (see Sect. 3.1) and form the actual solution approach to implement Heijunka in an assembly (Ohno, 2013). In its simplest form, this method places the different products as evenly as possible on the assembly line in a mixed-model assembly. An equal distribution of the products to the different shift programs in the production program planning is assumed. If the different products contain identical parts that are important for levelling the supply chain, this must be taken into account in level scheduling, which increases the complexity.

In larger companies, this task is usually performed by software; a very good overview of the numerous approaches to level scheduling can be found in the work of Boysen (2005, p. 217). A simplified approach, which is very often used in practice, focuses on a manual equal distribution of the different products and furthermore, in a second step, on an equal distribution of critical, predefined option variants.

Mixed-Model Sequencing

The goal of mixed-model sequencing is to minimize utilization losses in the form of model-mix losses (Sect. 2.8.2). This has been a very intensive branch of research for decades with a wide variety of approaches; a good overview is contained in the work of Boysen (2005, p. 219) and Huchzermeier and Moench (2022). Numerous software companies offer different solutions for mixed-model sequencing for large and small assemblies. The individual customer orders, based on the respective overloads and underloads per station, are sequenced to produce a balanced workload for each worker. Drifting of workers is minimized. In implementation, two different problems usually arise. Either the company, due to its overall size, has a functioning planning organization that can ensure the supply of basic data and the application of the mixed-model sequencing software. Then this company usually also has a very large production unit with many and very variant-rich orders (e.g., an automobile assembly)—the complexity of mixed-model sequencing is very much increased. Or it is a small assembly unit, the product volumes and options are significantly reduced, then these companies usually do not have the resources to perform mixed-model sequencing. Nevertheless, during the research for this book, we found that nearly all newly created assembly systems (with a focus on automotive production) will rely on mixed-model sequencing in the future.

Car Sequencing

Behind the idea of car sequencing is the desire to circumvent the data collection effort of mixed-model sequencing and, in addition, to generate a levelled flow of consumed material. It can thus be understood as a simple hybrid of the first two approaches. First, series succession restrictions are formulated on the product level and then series succession restrictions for defined option variants. So-called H_o:N_o-sequence rules are defined (Boysen, 2005). A practical example: out of 10 vehicles on the assembly line, only one convertible version may be present. Or, the distance between two units with a panoramic roof option must be two. Either manually or with the help of software, the assembly sequences are then formed from the shift programs of the production program planning.

Conclusion In the end, it must be noted that all three methods severely restrict—actually exclude—an orientation towards a true build-to-order (real customer demand is mapped 1-to-1). In existing assemblies, car sequencing is probably the most widely used concept currently. It tries to find the best compromise between a supply-oriented and overload-oriented approach. In the further course, we will show that the **need for the overload-oriented approach can be reduced by using**

variable takt—this is done by variable takt or VarioTakt. There can thus be a greater focus on the supply-oriented approach in series-sequencing. Or to put it another way: **the Heijunka task for the supply of material and preliminary products can be taken over by series-sequence planning, while the Heijunka task for a levelled workload on the workers, and thus a reduction in waste, is taken over by the VarioTakt** (see also Fig. 3.12). In industry, there is currently a clear trend toward increased use of the workload-oriented approach of mixed-model series-sequencing, for example in Factory 56 at Mercedes-Benz (Sect. 8.5). Increasingly better databases of the companies and the application of successful research work in software products for sequencing enable this development.

3.2.4 Heijunka with the VarioTakt: Levelling Despite Build-to-Order

Let's get a little ahead of our core Chap. 4 "Variable Takt" to illustrate its effectiveness as a method for levelling in mixed-model assembly of type 3. If previous production systems mainly use series-sequencing to implement Heijunka, the variable takt can help to better level an assembly line with different products in line balancing. In the following example, we illustrate the levelling effect of the VarioTakt compared to the fixed takt. We divide the fixed takt again into an application with closed station boundaries and a WATT application with open station boundaries. In all three examples, the sequence of orders on the assembly line is identical, two products with two mandatory option variants are manufactured, see Fig. 3.6. This results in a takt time spread of 6 time units (TU), which we can refer to as Mura.

In the following, we simulate the assembly sequence at a station with an identical sequence of products. In the assembly line with fixed takt (Fig. 3.7) and closed station boundaries, a takt of 11 TU is selected. In the figure, it is easy to see that the product P1a immediately generates an underload of 3 TU, due to the closed station boundaries this underload has a direct effect as waiting time, because the worker has to wait for the next product. Also, an overload, or Muri, on product P2a directly acts as Muda, since the worker cannot finish his/her work content. The line would be stopped, reassembly would be necessary, or a flexible worker would have to be used.

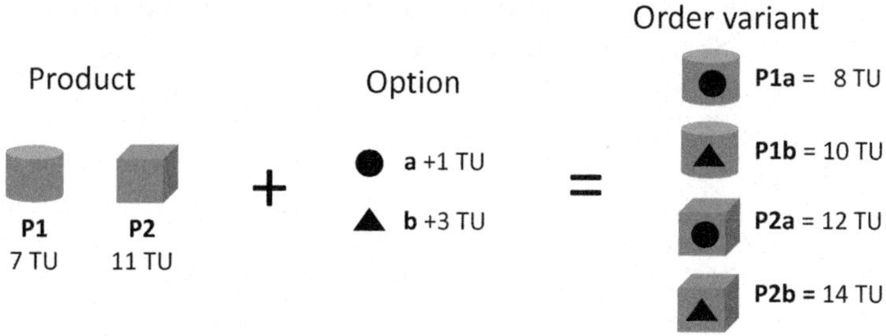

Fig. 3.6 Variant composition of products P1 and P2

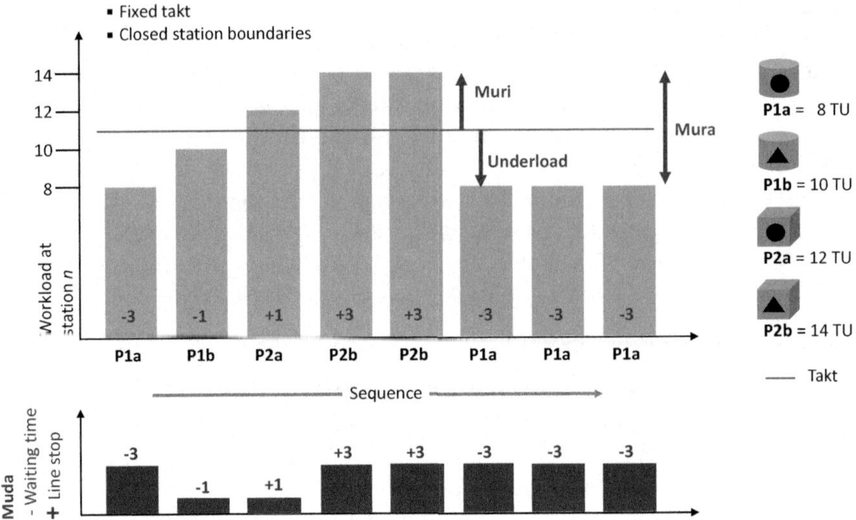

Fig. 3.7 Mixed-model assembly of type 3 with fixed takt and closed station boundaries

In our example (Fig. 3.7), in summary, there would be a loss of 20 TU at the station. And actually much more for the entire assembly line: a line stop immediately generates losses for all workers on the assembly line and thus transfers these losses to all stations. If the use of flexible workers is not possible and line stops should be excluded anyway, the takt could be set to 14 TU (the losses are fully realized as takt losses, see Fig. 3.4). In this way, all losses have an effect as waiting time and thus only in the respective station. However, an even larger loss of 30 TU in the form of waiting time occurs with a takt time of 14 TU.

By using the WATT method and open station boundaries, the losses can be reduced significantly, Fig. 3.8 illustrates this. Muri and underload have the same value for all products as in the previous example. The open station boundaries create a buffer, in our example +1 TU, which reduces the losses. If there is still compensation capacity in the buffer, overload and underload do not have an immediate effect as losses, i.e., they are mastered. The total loss is thus reduced to 15 TU. In this example, it can also be seen that the losses essentially occur in the form of model-mix losses (see Sect. 2.8.2) and are thus dependent on the sequence of products in the line. With the optimal sequence of the same products, such as P1b-P2a-P1a-P2b-P1a-P2b-P1a-P1a, the loss would be reduced to 7 TU.

With variable takt, the variance in an assembly line can be mastered even better. Figure 3.9 clearly shows the advantages of the VarioTakt, the combination of variable takt and the WATT method. We increase the number of takts to two and set the takt for all P1 products to 9 TU and that of P2 products to 13 TU. We describe the conceptual and practical implementation, advantages, but also prerequisites for the use of variable takt in detail in Chap. 4. By using the VarioTakt, we significantly reduce Muri and underload and thus also the total Mura. We achieve a better levelled workload of the stations, so that in our example there is only 1 TU of loss. However

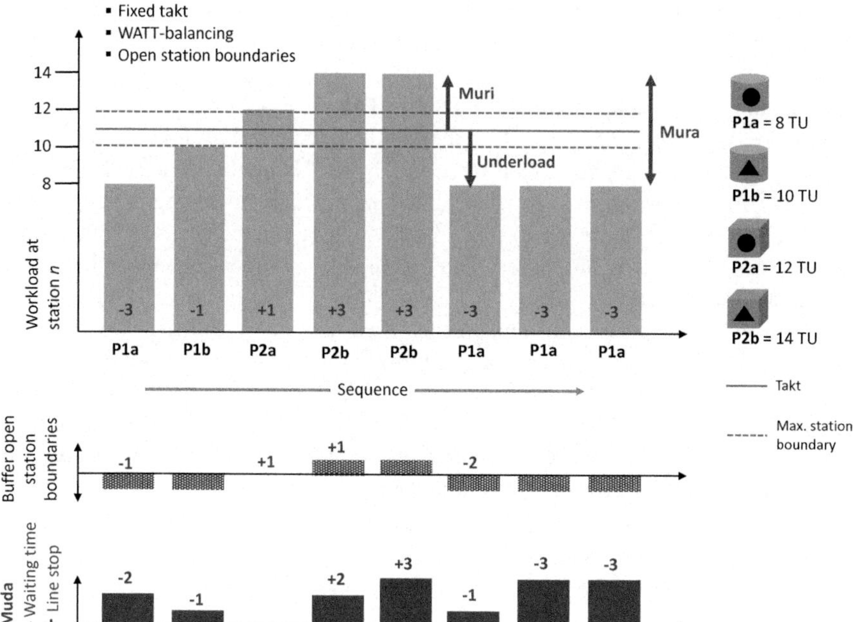

Fig. 3.8 Mixed-model assembly of type 3 with fixed takt and open station boundaries

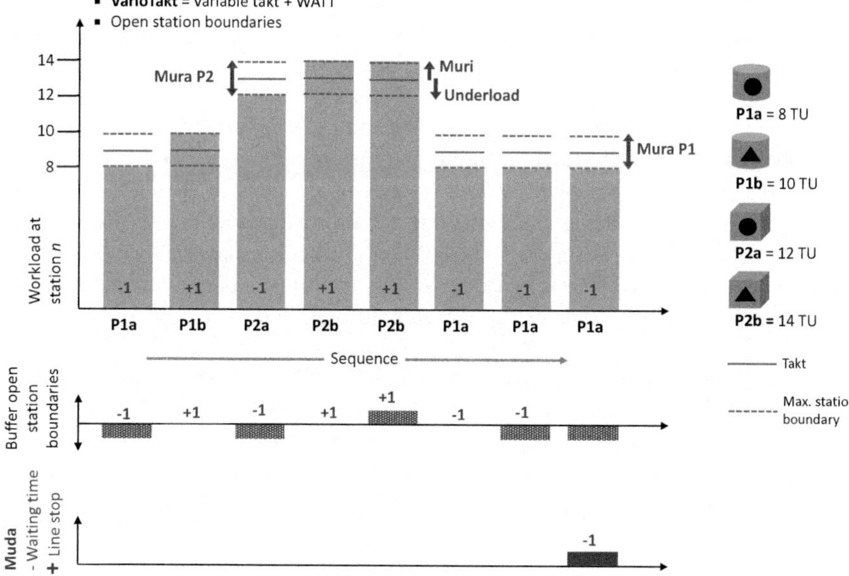

Fig. 3.9 Mixed-model assembly of type 3 in VarioTakt

the sequence of the jobs is chosen, the best results are always achieved with the VarioTakt.

Conclusion on Levelling with the Variable Takt

In Fig. 3.8, when using the WATT method in the fixed takt, we showed that with an improved sequence, the losses can be reduced again. Using the VarioTakt (Fig. 3.9), not only are these losses further reduced but there are many more degrees of freedom in designing the sequence. Or to put it briefly: There are many more combination possibilities with low or no losses. The complexity of series-sequence planning is significantly reduced.

When using the fixed takt, levelling is often sought by shifting the workload. Either workload peaks (work operations for options with a lot of work content) are shifted to pre-assemblies. Or gaps in the load of the takt (for products with less work content) are artificially filled with activities from pre-assembly. This shifting of work content is not necessary when using variable takt. **The assembly line levels itself in the variable takt—a built-in Heijunka effect** (see Sect. 4.5).

3.2.5 A Practical Example: Two Lines, Two Concepts

In Figs. 3.8 and 3.9 of the previous chapter, we presented two theoretical examples of balancing of a mixed-model assembly—the WATT method and the VarioTakt. In the following, we show two real examples using exactly these two principles and compare their effects. The data is taken from the work of Moench et al. (2020), which we will discuss in detail in Chap. 5.

The examples listed are from Fendt tractor production. To understand the data, it is important to know that the assembly of the tractor chassis and the tractor cabin (i.e., the driver's cabin) are independent of each other. This is the same across manufacturers, all tractors are built according to this basic structure. In most cases, the cabin is completely manufactured and assembled at a different location, transported to the chassis plant or chassis assembly line, and then placed on the chassis near the end of the assembly line. In the following example, we are dealing, on the one hand, with the cabin assembly line (without pre-assembly) and, on the other hand, with the first assembly section of the tractor chassis assembly (also without pre-assembly). In the cabin line, five different models of cabins are assembled, which are placed on the eight different models of tractor chassis. Thus, there are cabin models that are placed on different chassis. Both assembly lines employ a similar number of workers, but this is irrelevant for our example, since all data have been normalized to 100% and represent only a part of the assembly in order not to publish competition-relevant data. The data shows the ascending sorted assembly workload of all orders of a production week of the two assembly lines.

The decisive factor for comparing the two lines is that the variance of the total *assembly content* (line and pre-assemblies together) of the cabin and tractor body is approximately the same. Or, to put it another way, the difference in assembly workload between the smallest and largest cabin is the same as for the tractor chassis.

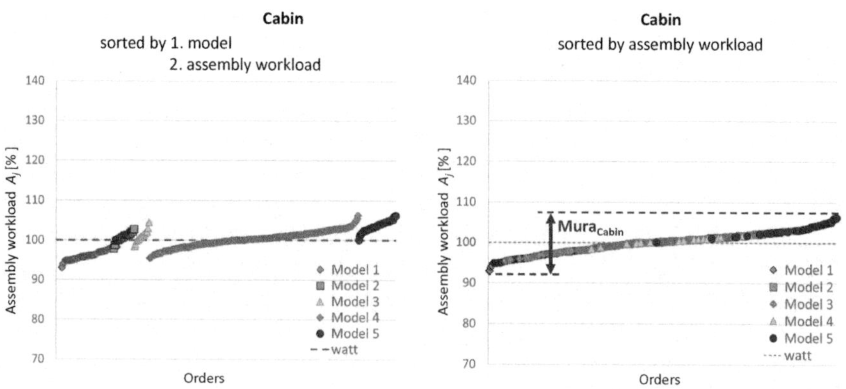

Fig. 3.10 Total assembly time of the main assembly line Fendt tractor cabin, sorted in ascending order by model and assembly time (left) and assembly time only (right) (adapted from Moench et al. 2020)

However, the following diagrams show a big difference, because the assembly line of the cabin is organized in the fixed takt, whereas the assembly line of the chassis uses the VariotTakt.

The total assembly workload of a cabin varies by 40%, but the assembly workload in the assembly line varies by about 15% (Fig. 3.10). The remaining 25% of the variation is mainly absorbed in pre-assembly. In order to reduce takt losses and model-mix losses, the assembly workload of the individual cabins and their order variants was adjusted over a longer period in the assembly line. Variant contents, especially of the more complex cabin models, were placed in pre-assembly and pre-assembly contents, of the less complex models, were pulled into the assembly line. This moving off or into pre-assemblies is not always efficient but serves the goal of reducing Mura in the line, i.e., Heijunka by using pre-assemblies. At first glance, this approach seems more successful in terms of Heijunka as in the tractor chassis assembly line (Fig. 3.11), and it is initially a noteworthy achievement of assembly planning. However, the variance has not disappeared, it is now concentrated on the pre-assemblies, which, decoupled by buffers, can handle variance much better than the assembly line in the fixed takt. But even this solution approach very quickly reaches its limits when there are strong fluctuations in the mix. Why not master the variance in the assembly line? The advantages of assembly in flow (Sect. 2.1) can be used in this way.

At first glance, the difference in workload for the workers in the line of the tractor chassis looks significantly greater than in the line of the cabin—and it would be even without variable takt. In this example, 8 takt times or takt time groups are used. Almost the entire variance of the workload of almost 60% is found in the assembly line. By using variable takt, the overloads and underloads do not result in takt or model-mix losses. *We will show how the assembly line self-adjusts to its workload, masters the variance and places levelled workloads on the assembly line at any given time.*

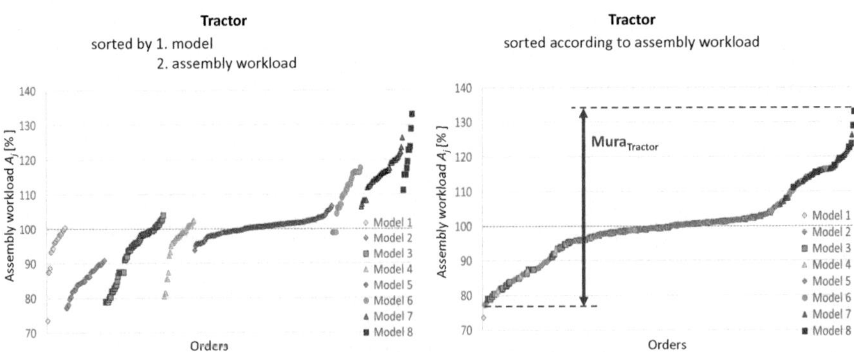

Fig. 3.11 Total assembly workload of the main assembly line Fendt tractor chassis, sorted in ascending order by model and assembly time (left) and assembly time only (right) (adapted from Moench et al. 2020)

3.3 Batch Production or Heijunka or Build-to-Order Afterall

If it is possible to map a lot of assembly content, and thus also option variance, in the assembly line itself and not have to relocate it to pre-assembly, all the advantages of clocked flow assembly, as we describe in Sect. 2.1, can be used.

- *Continuously flowing processes bring problems to the surface* (Sect. 2.1.1): In pre-assembly, problems often do not get through to supervisors, i.e., there is a disconnect. A flow assembly does not forgive mistakes, problems must and are solved sustainably.
- *Overproduction, and thus avoiding waste* (Sect. 2.1.2): In order to take advantage of temporary excess capacity or to protect against peaks in demand, buffers are often greatly extended in pre-assembly. Production defects are thus discovered later, more capital is tied up, more production space and load carriers are occupied, and damage is caused by more movement of the (pre-)products.
- *Aiming for perfection* (Sect. 2.1.3): In pre-assembly, poor work organization or (management) mistakes can usually be solved temporarily very easily with more personnel or buffers. In a clocked flow assembly line, this is not possible; more work content in the assembly line forces all those involved (management, planners, workers, etc.) to be much more disciplined and professional.
- *Line pressure* (Sect. 2.1.4): In Sect. 2.1.4 we described the ambivalent advantages of line pressure, it should seem obvious that these will not come into play in a pre-assembly.

Batch production (build-to-inventory) is certainly justified in many manufacturing processes for simple and low-variant products and is the most efficient way to manufacture these products. In the vast majority of assemblies, however, this is no longer the case. Increasing individualization of products rules out

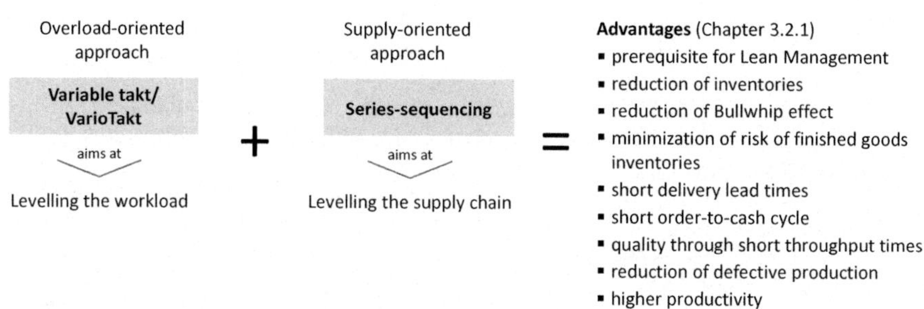

Fig. 3.12 Interaction between VarioTakt and series-sequence planning

batch production for assemblies. As shown in this chapter, the application of Heijunka according to TPS represents a compromise to achieve quality and efficiency advantages at the expense of inventories and longer delivery lead times. In the classic Heijunka approach according to TPS, a portion of waste is targeted to achieve production levelling, thereby reducing significantly more Muda than is used to achieve it. Variable takt can significantly assist in levelling an assembly line. Sequencing can be completely focused on its supply-oriented task. With different products, there is no need to artificially move pre-assembly content into or out of the assembly line. The assembly line, and thus the workforce need, levels itself autonomously (Fig. 3.12).

Our focus is on the use of variable takt in the assembly planning phase, in AD for MMA—in particular in the line balancing of the assembly line. At the end of our book in Sect. 8.6.4, we additionally outline a concept to use variable takt also in SD for MMA, more precisely in overload-oriented series-sequence planning. Thus, variable takt can not only level the global variance (Sect. 2.4.2) between the products but also compensate for temporary and local workload peaks, through the mix of options variants.

With the VarioTakt it is possible to come very close to a *real* build-to-order in the assembly line. This includes the fastest possible delivery lead time as well as unlimited configurations of products! New technologies, applications from the field of Industry 4.0 and the further development of different production concepts (Chaps. 7 and 8) will enable assembly lines to develop more and more in the direction of a true build-to-order—the necessity and the will exist!

References

Boysen, N. (2005). *Variantenfließfertigung* (294 p.). Springer. https://www.springer.com/gp/book/9783835000582

Chen, F., Drezner, Z., Ryan, J. K., & Simchi- Levi, D. (2000). Quantifying the bullwhip effect in a simple supply chain: The impact of forecasting, lead times and information. *Management Science, 46*(3), 436–443. https://doi.org/10.1287/mnsc.46.3.436.12069

Da Silveira, G., Borenstein, D., Fogliatto, S., & F. (2001). Mass customization: Literature review and research directions. *International Journal of Production Economics, 72*, 13. https://doi.org/10.1016/S0925-5273(00)00079-7

Diwas, S. K., & Terwiesch, C. (2009). Impact of workload on service time and patient safety: An econometric analysis of hospital operations. *Management Science, 55*(9), 1486–1498. https://doi.org/10.1287/mnsc.1090.1037

Domschke, W., Scholl, A., & Voß, S. (1997). *Produktionsplanung: Ablauforganisatorische Aspekte* (455 p.). Springer.

Fujimoto, T. (1999). *Evolution of manufacturing systems at Toyota* (400 p.). Taylor & Francis.

Huchzermeier, A., & Moench, T. (2022). Mixed-model assembly lines with variable takt and open stations: The ideal and general cases. *Unpublished working paper. WHU – Otto Beisheim School of Management, 20*(3), 35 p.

Huchzermeier, A., Moench, T., & Bebersdorf, P. (2020). The Fendt VarioTakt: Revolutionizing mixed-model assembly line production – Case article. *INFORMS Transaction on Education, 20*(3), 125–176. https://doi.org/10.1287/ited.2019.0224ca

Kilbridge, M., & Wester, L. (1966). An economic model for division of labor. *Management Science, 12*(6), 255–269. https://doi.org/10.1287/mnsc.12.6.B255

Kotha, S. (1995). Mass customization: Implementing the emerging paradigm for competitive advantage. *Strategic Management Journal, 16*, 21–42. https://doi.org/10.1002/smj.4250160916

Lee, B. H., & Jo, H. J. (2007). The mutation of the Toyota production system: Adapting the TPS at Hyundai motor company. *International Journal of Production Research, 45*(16), 3665–3679. https://doi.org/10.1080/00207540701223493

Lee, H. L., Padmanabhan, V., & Whang, S. (1997). Information distortion in a supply chain: The bullwhip effect. *Management Science, 50*(12), 1763–1893. https://doi.org/10.1287/mnsc.1040.0266

Liker, J. K. (2004). *The Toyota Way* (330 p.). McGraw Hill

Liker, J. K. (2021). *The Toyota Way* (449 p., 2nd ed.). McGraw Hill.

Moench, T., Huchzermeier, A., & Bebersdorf, P. (2020). Variable takt time groups and workload equilibrium. *International Journal of Production Research*. https://doi.org/10.1080/00207543.2020.1864836

Ohno, T. (2013). *The Toyota production system* (176 p.). Campus Verlag.

Pine, B. J., Bart, V., & Boynton, A. C. (1993). Making mass customization work. *Harvard Business Review* Online. Accessed April 29, 2021, from https://hbr.org/1993/09/making-mass-customization-work

Rother, M. (2009). *Toyota Kata* (400 p.). McGraw-Hill.

Rother, M. (2013). *Die Kata des Weltmarktführers: Toyotas Erfolgsmethoden* (299 p.). Campus Verlag.

Swist, M. (2014). *Taktverlustprävention in der integrierten Produkt- und Prozessplanung* (328 p.). Apprimus Verlag.

Thomopoulos, N. T. (1966). *A sequencing procedure for a multi-model assembly-line*. Illinois Institute of Technology.

Thomopoulos, N. T. (1967). Line balancing-sequencing for mixed-model assembly. *Management Science, 14*(2), B-59-B-75. https://doi.org/10.1287/mnsc.14.2.B59

Veit, M. (2010). *Models and methods for inventory sizing in Heijunka-leveled supply chains* (129 p.). Faculty of Mechanical Engineering, Karlsruhe Institute of Technology (KIT).

Womack, J. P., Jones, D. T., & Roos, D. (1990). *The machine that changed the world* (323 p.). Westview Press.

4 The Variable Takt

We come to the core of our book. In our Mixed-Model Assembly Design (MMAD, Sect. 1.4), it is our most important element from Assembly Design for Mixed-Model Assembly (AD for MMA). As in the other chapters, we try to stay close to the language of Operations Management practice in the form of presentation, but further see ourselves as mediators of current findings from scientific work. For motivation, we comment again in the first part on an increased need for mixed-model assembly for current and future production sites, which we had already touched on in the introduction to this book. This is followed by a brief explanation of the classic solution strategies for coping with variants in assemblies with fixed takts, because the disadvantages of these can be eliminated with variable takt. In Sect. 4.4, we explain in detail how variable takt works, with its effect on takt times and the sizing of assembly lines. In Sect. 4.5, we show the numerous advantages of variable takt, using many practical examples. We explain the necessary technical requirements for implementing variable takt in Sect. 4.6 and the possibilities for linking different assembly sections in Sect. 4.7. This chapter concludes with the experiences gained from more than 4 years of practical implementation of variable takt in tractor assembly at Fendt.

The explanations on the division of the product portfolio into takt time groups are elementary for all those who are willing to implement variable takt; these are given in Chap. 5.

As Fig. 4.1 shows, the goal of AD for MMA concepts is to minimize takt losses, ideally up to the boundary curve in the MMAD solution space. We have already introduced different concepts in structure, organization and balancing in Chap. 2 and summarized them in Sect. 2.9. With this chapter, we now add the variable takt as one of the most powerful concepts to master variance in mixed-model assembly. For a simplified understanding of variable takt, we always assume a mixed-model assembly line of type 1 in this chapter. This means that the option differences of products do not play a role at first. As already described in Chap. 2, variable takt and the WATT balancing method can be combined to form the VarioTakt in order to

P. Bebersdorf, A. Huchzermeier, *Variable Takt Principle*, Management for Professionals, https://doi.org/10.1007/978-3-030-87170-3_4

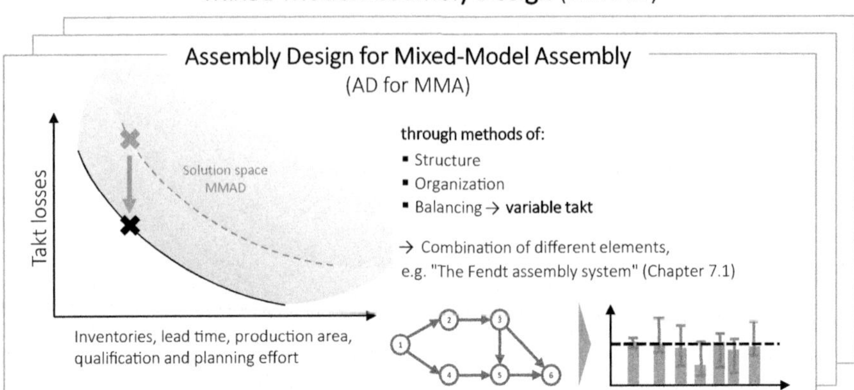

Fig. 4.1 Assembly design for mixed-model assembly in MMAD

assemble different products with different options in a type 3 mixed-model assembly line.

4.1 The Need for Mixed-Model Assembly

Among the classic, sometimes already disruptive, challenges for the conceptualization and management of today's industrial production are three major trends that Huchzermeier and Moench (2022) see confirmed by many contributions to the INSEAD-WHU Industrial Excellence Award (IEA, formerly "The Best Factory").

- *Increasing product individualization and centricity* (Sect. 1.2.1)
  Ever more individual product derivatives, specially adapted to the most diverse customer requirements, have to be manufactured—from mass production to mass customization to mass individualization.
- *Localization to hedge global uncertainties* (Sect. 1.2.2)
  The trend toward large mega-factories and the hoped-for exploitation of economies of scale is no longer recognized as being effective. Risks must be distributed. A digital, but above all standardized assembly for customer-specific products with a fixed and thus limited number of units as a blueprint for further globally distributed factories is coming to the fore.
- *Disruptive innovations and technology diversification* (Sect. 1.2.3)
  New technologies or features must be integrated into existing traditional fabrication and assembly processes. The associated differences in assembly workload and an increasing takt time spread must be mastered.

The strongly increasing part variance can be solved with logistics concepts from Lean Management (JIT/JIS, Kanban, etc.), increasing digitalization and automation.

Late variant creation, made possible by the design of new products, will also help to solve this challenge, but this topic should not be our focus.

With innovative technologies and intelligent fixture concepts, i.e., classic strengths of German machinery and plant engineering, the variances in cubature and weight of the different products and options can be solved.

Increasing product individualization, technology diversification and the localization of production mean that a wide variety of products have to be manufactured on a production line, especially in the area of assembly. This in turn results in requirements for assemblies that can initially be perceived as a contradiction to the classic assembly line.

Assemblies must be able to map:

1. variations in mix and volume
2. for different products
3. with high spread in the basic assembly times
4. in a wide range of option variants
5. in continuous flow
6. in total-mix, according to build-to-order logic

4.1.1 The Need for Mixed-Model Assembly Using the Example of Fendt Tractor Assembly

Even before these developments, there are and were branches of industry that have been forced to produce their very different products on one assembly line for decades. A very good example is agricultural machinery and here the tractor production of the Fendt brand, which we described in detail in the introductory Sect. 1.3. A tractor assembly is characterized by four essential elements:

First—low total number of units vs. many model series A rather low number (compared to the automotive sector) of units sold contrasts with a large number of very different model series. In the following chart (Fig. 4.2), we normalize the number of units produced by Fendt to a volume of 10,000 units; the differences in volume between the individual model series are extreme. For example, the share of the production volume of the top model Fendt 1000 Vario is about 5%, which corresponds to only a few units on one production day. Despite this small number of units, this represents a global market share for this type of tractor of over 50%. A challenging application for a classic flow assembly line.

Second—extremely different dimensions The different tractors differ considerably in their dimensions and cubature (see Fig. 4.3). For example, only the transmission of the Fendt 1000 Vario has a weight of 3.3 tons, while the complete Fendt 200 Vario tractor weighs 3.9 tons. The Fendt 200 Vario could be completely covered by the engine hood of the Fendt 1000 Vario. Despite these extreme differences in design and consequently overall work content, these products are assembled on only

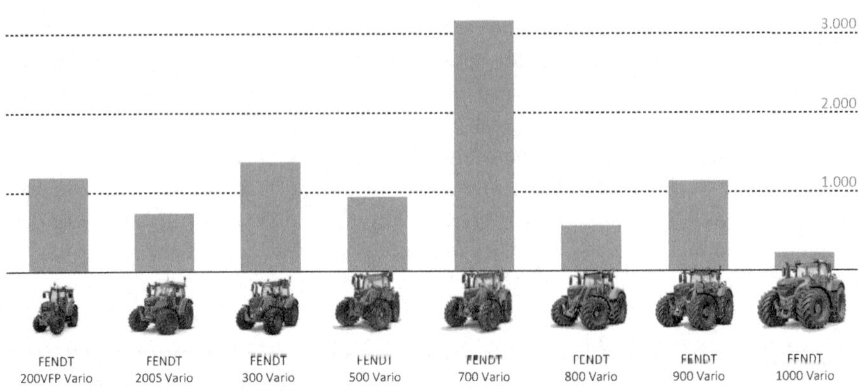

Fig. 4.2 Unit number distribution Fendt—normalized to 10,000 units

Fig. 4.3 Total-mix production extreme—the Fendt 1000 Vario (left) and the Fendt 200 Vario (right)

one assembly line in total-mix. Independent assembly lines for these products cannot be operated economically. Due to the seasonal sales differences between the tractor models, it is also impossible to distribute the models on two or three assembly lines. The volume fluctuations of the individual models are levelled out much better on a common assembly line.

Thirdly—different technology structure Due to a very wide range of applications for tractors, from vineyard tractors to small farm tractors for dairy farms, professional machines for contractors or construction site vehicles, there are a large number of different tractor models that represent independent series in terms of design. Taking Fendt as an example, there are eight independent series on one assembly line. Here, the difference is not only between a "small" tractor and a "large" tractor, the differences concern the basic design of the tractors. In some types, the engine is a load-bearing element, while in others it sits in the base frame of the tractor. Braking and steering systems differ in their basic function between hydraulic and pneumatic—to name just a few examples of the design differences.

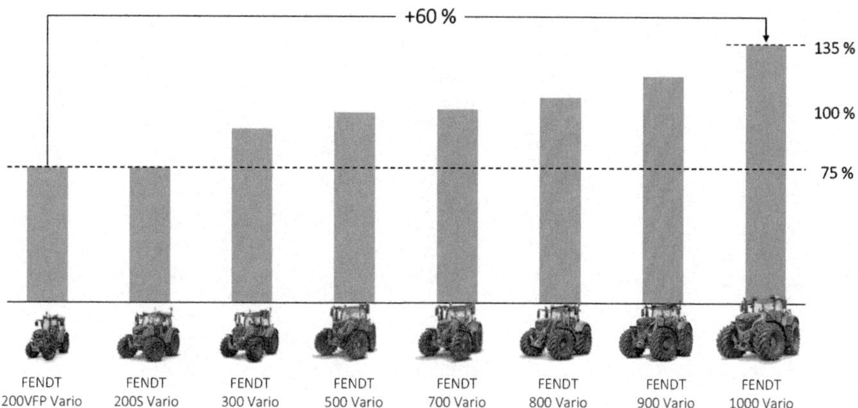

Fig. 4.4 Workload differences in assembly of the 8 series of the Fendt tractor portfolio (updated and adapted from Huchzermeier & Moench, 2022)

Fourth—different workload As already described, not only the variety of parts and different product dimensions put an assembly line to the test, it is mainly the big difference in total assembly workload. In Fig. 4.4, differences in the total work content of the assembly of different Fendt tractors are shown.

In summary, a tractor assembly line faces the challenge of effectively and efficiently assembling different products with high process time spreads in a flowing total-mix production.

4.2 Solution Alternatives with Fixed Takt

Before we get to the detailed description of variable takt, we will show the three most classic solution alternatives for handling different products with a fixed takt and the challenges that arise in an assembly. If the goal in an assembly line with a fixed takt is to produce products with different total workloads, the question arises: In which takt time? (Fig. 4.5)

In the literature, the fixed takt is also referred to as FRL, fixed-rate launching (Thomopoulos, 1967; Dar-El, 1978), or more simply, units are continuously placed on the assembly line at or after a fixed time interval, independent of the product itself. In the further course of assembly, the units are moved on an assembly line at a constant speed. This results in two basic problems that have already received a lot of attention in the scientific literature: the assembly line balancing problem, i.e., distributing the work content as evenly as possible to each station (Boysen et al., 2007), and the assembly line sequencing problem, i.e., finding the most ideal sequence of the different orders on the assembly line, taking into account series-sequence constraints that must be strictly observed (Thomopoulos, 1967). We will

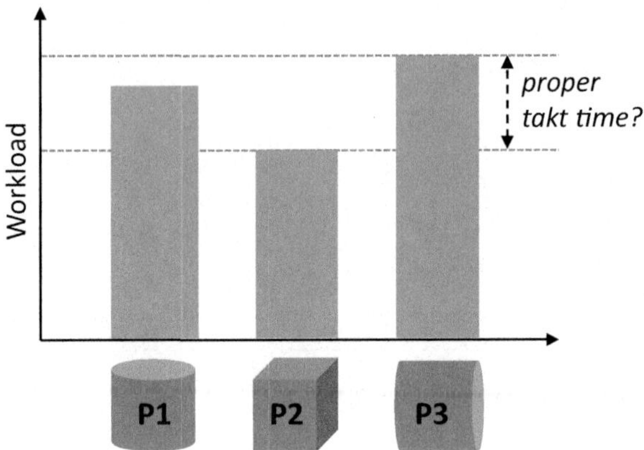

Fig. 4.5 Which fixed takt time should be selected?

show *that variable takt can significantly simplify the balancing problem and the series-sequencing problem!*

One of the biggest bottlenecks in mixed-model assembly is series-sequence restrictions, which are very often due to the balancing of workload in the fixed takt. This means that customer demand (build-to-order) cannot be produced exactly, or only with a time offset (Huchzermeier & Moench, 2022). In times of rapidly changing and uncertain markets, these series-sequencing constraints create a strong inflexibility.

If a fixed takt is adhered to, there are three classic and in practice widespread solution options for assembling different products with a different total assembly workload. As already described in the introduction to this chapter, in order to simplify the description of the principle of operation, we will assume in the following a mixed-model assembly of type 1 (cf. Sect. 2.5), i.e., with different products, but no options (Fig. 4.6).

These "classics" have proven themselves for many practical applications. Parts of them are taken up again in Chaps. 7 and 8. However, in many cases these alternative solutions are poor compromises:

- *Maximum takt*
  This variant is certainly the simplest solution to implement, but at the same time also the one with the least required management know-how. It is functional, is completely independent of the model-mix and does not lead to an overload of workers at any time. All losses occur in the form of takt losses, model-mix losses are excluded. However, the disadvantages are also obvious—there are huge efficiency losses. The greater the workload difference between products and the greater the number of products with small workloads, the greater the takt losses. Since the takt time is oriented to the product with the largest total assembly workload, the total volume of assembly is reduced, and capacity reserves cannot be used.

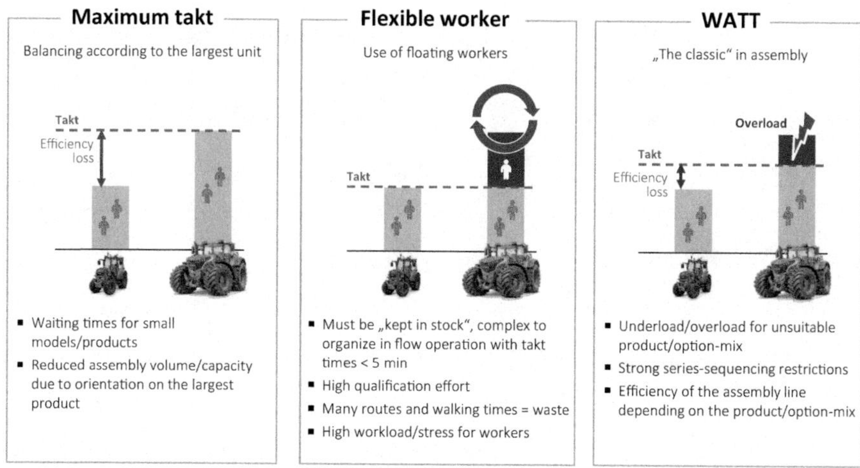

Fig. 4.6 The classics—dealing with different takt times using the example of Fendt tractor assembly

- *Flexible workers*
  The use of flexible workers is also widespread. A distinction can be made between two implementation variants. The use of flexible workers, which is planned in advance, can be "migrated" from pre-assembly to the assembly line in the event of a defined overload or can be permanently located on the assembly line. A second option is not to plan the flexible worker activities in advance and to leave it up to the system to call in flexible workers as needed. However, the smaller the takt times, the more difficult it is to plan this process. The workload and qualification requirements for these employees are considerable. Flexible workers can never be used efficiently; there are losses due to walking distances, workload imbalance and idle times. In addition, this variant assumes that the model-mix meets firmly defined criteria. Too many products with an increased workload in succession bring the assembly line and above all the employees to their workload limits, a stop of the assembly line is then usually unavoidable.
- *WATT balancing*
  With WATT (weighted average takt time), the entire line is designed for an average takt. A detailed description of this takt method can be found in Sect. 2.6. The WATT is the classic method in the automotive industry. On the one hand, decisive for a successful functional implementation of the WATT is the maximum permissible overtakt, the drift ranges and the associated restrictions in series-sequence planning so as not to overload the assembly line. Decisive for productivity, on the other hand, are the takt losses for small types (individual workload smaller than the average takt time) and the overloading of employees for types with a workload above the takt time (Muri), because here there is an increased risk of line stops or the increased use of flexible workers. Consequently,

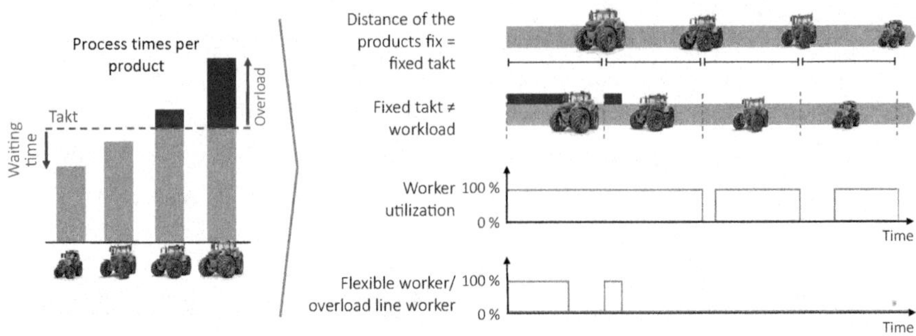

Fig. 4.7 Effects of fixed-takt mix production using the example of Fendt tractor assembly (adapted from Moench et al., 2020a)

there is a strong dependence on the model-mix of the assembly line, strong series-sequence restrictions are the consequence.

These disadvantages of the fixed takt were summarized very well by Moench et al. (2020a) in Fig. 4.7.

We have already pointed out several times that it is almost impossible to create a continuous and levelled flow of workload when assembling different products in a fixed takt. In this chapter of our book, we now reveal the secret about how variable takt works—it is easy, even simpler, than we thought.

4.3 Differentiation from Variable Rate Launching

In order to define the term variable takt more precisely, we must first deal with the term variable rate launching (VRL) or "variable launching intervals". In literature and science, the term VRL has been used since 1962 by Kilbridge and Wester. Variable-rate launching can therefore not really be called an innovation; the first approaches have been described in the literature since 1963. But what is a good idea without implementation! In numerous other works in the following decades, by Dar-El (1978), Bard et al. (1992), Boysen et al. (2007, 2009a), Fattahi and Salehi (2009), Tong et al. (2013), to name a few, the term is further coined. However, in the rest of our book, we do not want to equate this term with variable takt.

A first more detailed definition of VRL is given by Dar-El (1978). In his work, however, it can be seen that he uses VRL differently to our understanding of the variable takt. He describes a mixed-model assembly line by means of four dimensions, one of which he calls "launching discipline" (Dar-El 1978, p. 314). Implicit in this designation is the assumption that there is also an "undisciplined" launching of products onto the assembly line. For him it is the VRL, because "on completing work on current product, the first stations operator can immediately begin working on the next product" (Dar-El, 1978, p. 314). Thus, the first worker on the line specifies the run interval, the launching rate—without specifying a takt time.

If this insight is combined with its second dimension, the possibility that workers can temporarily detach the assembly object from the assembly line (Dar-El, 1978), this is not compatible with our understanding of variable takt. All subsequent publications do not clearly refer to this or insufficiently describe this property of VRL. For this reason, we define the term variable takt more precisely.

In order to use the variable takt as we understand it, the underlying assembly system must meet the following requirements:

1. Different, but firmly defined launching intervals, and thus
2. fixed takt times within the takt time group (see Chap. 5),
3. a continuous and constant (conveyor) speed of the assembly line,
4. no buffers between the stations, and thus
5. no changes in the series-sequence, the series-sequence is identical at each station.

4.3.1 Origin of the Designation "VarioTakt"

We refer to the combination of variable takt and the WATT takt as VarioTakt. The designation was created during the implementation of the variable takt in the tractor assembly of Fendt at the Marktoberdorf site and this in 2015 without knowledge of the descriptions of the variable takt or VRL from the scientific literature. Since 2012, Fendt assembled its tractors in a fixed takt on a total-mix assembly line. The tractors were moved in a continuous flow on the line with the help of autonomously driving AGVs (automated guided vehicles). Fendt used the principles from Chap. 2 to compensate for the differences in its tractor models when it comes to balancing. The introduction of the new Fendt 1000 Vario broke all previously known dimensions for assembly. Larger and heavier components could be accommodated by modifying the technology, but this new tractor model had significantly more work content than any other tractor. If Fendt had also integrated this type of tractor into the assembly line with a fixed takt, the utilization losses, in particular the takt losses, would have increased significantly. So, it came about that some experienced planners and managers together came up with a short and simple idea: "just change the distances". The basis for this idea was the experience Fendt had gained in batch production of assembly in the apron conveyor before the year 2012. Details on this are described in Sect. 4.6 in the section "Variable takt in the apron conveyor". Fendt was already implementing a kind of variable takt at this time, without calling it that.

But now moving away from the fixed takt? For the management at the time, this was not at all in line with the principles of Lean Management (Womack et al., 1990; Ohno, 2013), which adhere so strongly to the fixed takt—change anyhow? However, the idea was discussed further and the adherence to the fixed takt was fundamentally questioned, because on closer examination and with a deep understanding of the basic ideas of Lean Manufacturing principles (Ohno, 2013; Liker, 2021), it is not the fixed takt that is crucial. What is critical for effective (no defects) and efficient (no losses) operation of an assembly line is, first and foremost, a balanced and even workload for the workers. A fact that every experienced production manager immediately understands. In further discussions and simulations, the management realized that this is exactly where the advantages of variable takt lie. After this realization, the successful implementation of VarioTakt still had to overcome a number of technical and organizational hurdles, but to this day it remains one of the most decisive reasons for the success of Fendt brand tractor production. The name Vario was chosen in reference to Fendt's Vario transmission. The Vario transmission is one of the most important unique selling points of Fendt and one of the greatest innovations in agricultural technology in recent decades. It is a mechanical-hydraulic power-split transmission. It is the most important component for getting the tractor's drive energy "up and into the ground" in a targeted manner and without shift stages. It can provide the energy very variably from 0.1 km/h to 60 km/h speed, always in line with the application, with minimal losses—a fitting parallel to the variable takt and the WATT method, hence: the VarioTakt.

4.4 The Variable Takt for Mixed-Model Assembly

It is generally known that innovations are often slowed down or even prevented by sticking to existing systems, structures and thought worlds. Perhaps this is one of the reasons why the overwhelming number of managers and production planners, despite major differences in the workload of their products, stick to a fixed takt.

Wouldn't it be great if, in a clocked flow assembly, each product (type) had its own suitable takt? Maybe even not just every product, but every individual order? If the assembly lines and employees were level-loaded at all times, regardless of the product mix? What seems unrealistic and unfeasible for the majority of today's production managers in a variant-rich flow assembly is possible with variable takt.

For a simplified explanation of variable takt, we again use a mixed-model assembly of type 1 with products of different total workloads, but without option variants. In an implementation in a mixed-model assembly of type 3, the variable takt is combined with the WATT method to form the VarioTakt.

4.4.1 How the Variable Takt Works

Let us explain how the variable takt works on an assembly line with three different products. In our example, the assembly line has $N = 100$ stations, one worker w per station and our product 1 (P1) has a total assembly time of $A_{P1} = 500$ min. In Sect. 4.4.3, we explained the calculation of the takt time; to keep our example simple we assume we could distribute this assembly time ideally over the 100 stations. This results in a takt time $T_{P1} = 5$ min. With a station length of 5 m, this results in an assembly line speed of 1 m per min or $V_{CS} = 1.67$ cm per second.

The following illustration (Fig. 4.8) describes the sequence of a takt in the worker triangle (Sect. 2.3.1). The worker starts at the end of the station and collects information and material for the next job on his/her way back to the start of the station, against the direction of flow of the assembly line (note: this time expenditure

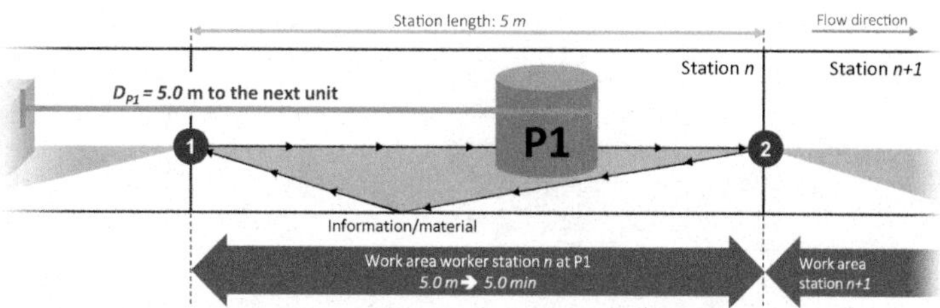

Fig. 4.8 Assembly sequence product 1—a classic sequence in the worker triangle

must be included in the 5 min of work content). He/she starts at position 1 with his/her activity on the product, moves in the direction of flow with the product while performing his/her activities. After 5 min or 5 m, he/she reaches the end of the station, he/she finishes his/her work at position 2. At this time, the next job also reaches the starting point of his/her station, a classic assembly sequence.

Note: We always assume that the worker starts working on the assembly object as soon as it enters his/her station. He/she does not have to wait until it is completely (with its entire cubature) in the station. The worker finishes his/her activity as soon as the assembly object reaches the rear station boundary (it enters the next station), the entire cubature of the assembly object has not yet exceeded the station boundary at this point. In the end, it does not matter what time is chosen to start the assembly, the variable takt works with any definition.

Product 2 (P2) has a 20% lower work content than P1, so $A_{\text{P2}} = 400$ min. If we had a separate assembly line with $N = 100$ stations for this product, the takt time would be 4 min.

If we want to mount it on the same assembly line as P1 in a fixed 5-min takt and apply the maximum takt (see Fig. 4.6), there would immediately be a 20% productivity loss due to idle times at P2. If the WATT were to be used, series-sequence restrictions immediately arise and model-mix losses become possible.

With variable takt, we can solve these problems completely. Since P2 has only 20% of the assembly content of P1, we also set the gap 20% shorter. In short, the gap behind P2 is thus only 4 m, $D_{\text{P2}} = 4$ m. In the context of VRL (variable rate launching), this means that a new unit is launched 4 min after each P2, regardless of the type of this subsequent unit. Of course, the speed of the assembly line remains identical at 1.67 cm/s.

Figure 4.9 again illustrates the sequence in the worker triangle. The worker starts at the end of the station, collects information and material on the way back to the beginning of his/her station. He/she starts his/her activity at position 1 and finishes it already after 4 m or 4 min at position 2 (see Fig. 4.9)—more work content is not assigned to him/her in the SWS. He/she must also have finished his/her activity at this position, because the next job (regardless of whether P1, P2 or P3) is already at the entrance to his/her station. In the last section, of 1 m length, in station n, no work is performed on P2.

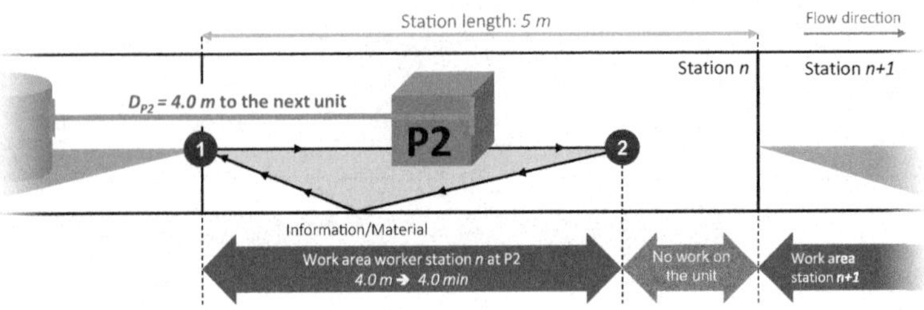

Fig. 4.9 Assembly sequence product 2—assembly line is virtually shortened

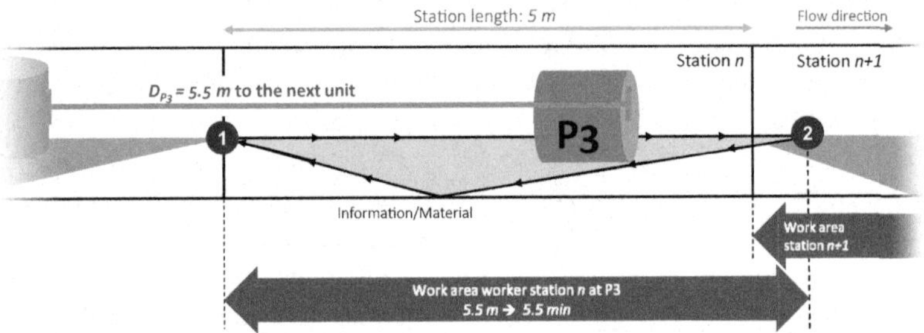

Fig. 4.10 Assembly sequence product 3—assembly line is virtually extended

Product 3 (P3) has a 10% increase in total work content, $A_{P3} = 550$ min. There is 5.5 min of work content per worker. The distance for this unit is therefore set to $D_{P3} = 5.5$ m. The assembly line continues to maintain its speed of $V_{CS} = 1.67$ cm/s.

In order to evenly balance the line with products P1, P2, P3 at a fixed takt with the WATT balancing method, i.e., to implement Heijunka, the repetition rate of P1 is not decisive. However, P2 and P3 would have to be levelled distributed in the series-sequence with a ratio of 1:2; otherwise, model-mix losses would occur in the form of idle times or the workers would drift off successively, which would lead to a stop of the line (Sect. 2.8.2).

Figure 4.10 describes the procedure for assembling P3. As always, the worker starts his/her work at position 1. Since the worker now has 5.5 min of work to complete, he/she flows 0.5 m into the following station as planned. In contrast to the fixed takt, this is not problematic, since a new order (regardless of whether P1, P2 or P3) only enters his/her station at the time when the worker is at position 2. He/she therefore still reaches the following unit at the start of his/her station in time. *No* drifting takes place.

It should be noted in this case that two workers are working on the product at the same time on the 0.5 m in which the worker is located in the subsequent station $n + 1$.

The size of the product and the work distribution on the product must allow this or must be planned in this way. In practice, this is often easy to implement, since the sequence of the various activities is oriented along the product in the direction of flow if the work operations are well balanced in that station. In short: the worker starts with activities at the front of the product (in flow direction), followed by further work content until he/she reaches the rear part of the product. If parallel work is not possible, it would also be possible in our example to set the station length to 5.5 m, the principle of the variable takt would remain, the total length of the line would increase by 10%. A detailed consideration of this topic can be found in Sect. 4.4.4.

And what about drifting?
When dealing with the variable takt for the first time, the impression very often arises that the workers could drift off continuously. In practice, drifting is defined as the continuous shifting of the work starting point of the workers due to excessive underload and/or overload. This drifting occurs when the station boundaries are open in the WATT balancing. If there are too many units with low workloads, it will be in the opposite direction to the flow direction of the line; if there are too many units with high workloads, it will be in the flow direction of the line. This will not happen with variable takt. Consider again the illustrations of the work sequences of products P1, P2 and P3. Let us assume that the sequence consists of an infinite number of products P2; in this case, the worker will always finish his/her work before the end of his/her station, he/she has to in this case, because the distance behind his/her order P2 is smaller than his/her station length (in our example 20% smaller). Thus, he/she always starts at the beginning of his/her station and does not drift against the flow direction. In this scenario, there are 20% more units in total in the assembly line. With P3 it is exactly the other way round—he/she flows with product P3 approx. 0.5 m into the next station, but exactly by this amount the distance behind P3 is larger, thus the worker reaches the next order exactly at the beginning of his/her station again. In total, there are 10% fewer units in the assembly line in this scenario. From this description, the difference to drifting, as we have described it in Sect. 2.8.2, should have become clear.

Minimum distance and line speed
The distance of the units is always set from the "tip" of the mounting object to the "tip" of the following object. It should be noted that, depending on local conditions, each assembly object must still be followed by a safety distance. It follows that the distance of the assembly object with the smallest variable takt time must at least correspond to the length of the assembly object and the safety distance. With our definitions of the speed of the assembly line in the following Sect. 4.4.3, this results in the minimum speed of the assembly line.

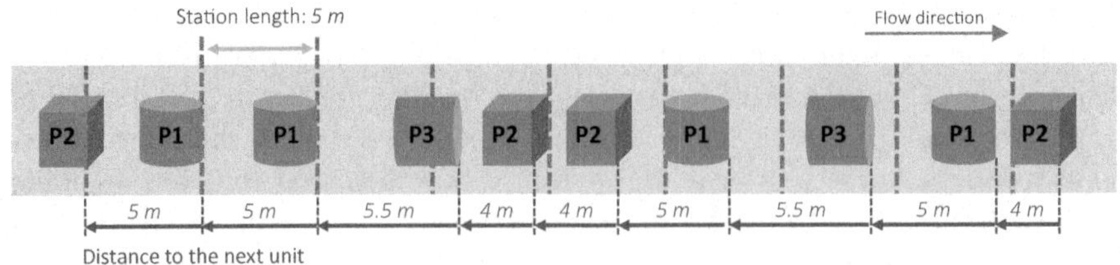

Fig. 4.11 Assembly line occupancy with variable takt—an independent heijunka

4.4.2 Levelling with Variable Takt

If we consider any number of stations with a random mix of our three products, the picture in Fig. 4.11 might emerge.

At first glance, this assembly line appears uneven, unbalanced. However, this impression only arises if we cast our eyes on the different distances. If we direct our gaze to the distribution of the workload, a completely different picture emerges. In this view, the number of products leaving the line in a fixed time interval is indeed different. However, *the workload leaving the line is almost constant per time interval.* Since products with a higher workload usually also generate more sales, it could also be said that this assembly line generates a more constant output of sales, regardless of the mix.

If there are more products with a low workload in the line, there will be more products in total in the line. If there are more products with a high workload in the line, then there are fewer units in total in the line. In other words, *the line balances itself automatically, regardless of the model-mix!* Therefore, no computer-aided sequence optimization is needed to find the best compromise in series-sequencing—the assembly line is always evenly loaded. This is perfect levelling—Heijunka!

Regardless of the model-mix, this type 1 mixed-model assembly line never creates idle times that result in lost productivity or overloads that lead to workers continuously drifting off.

From a broader perspective, it could also be said that *the assembly line is virtually shortened for products with low work content. For products with larger work content, it is virtually lengthened.*

If the findings of this chapter are drawn together, it can also be seen that the fixed takt, is ultimately only a special subtype of the variable takt, where the "distance" is constant, i.e., fixed (Huchzermeier & Moench, 2022).

4.4.3 Determining Variable Takt Time

We have not yet answered one relevant question. What is the appropriate takt time for a product in variable takt? In our very simplified example from this chapter, this question is easy to answer, it lies in the relation of the total assembly workload and the number of workers in the assembly line. It should be emphasized that in mixed-model assembly with variable takt, the number of workers must always be kept the same for each product; otherwise, losses are incurred due to waiting or idle time.

$$Takt\ time = \frac{Total\ workload}{Number\ of\ workers}$$

Admittedly, this formula is very simplified. It would correspond to a 100% balancing efficiency, which would never be feasible in practical applications. Work operations (WOs) cannot be divided arbitrarily small to fill all T-bar charts to 100%.

*Note: All the following explanations, calculations and examples are based on a mixed-model assembly line of type 1. Different products without option variants are assembled on the assembly line. This means that the assembly content of each customer-specific order of a product is the same (*cf. Sect. 2.4.1*) or, in short: all orders of product P1 have the same assembly workload* A_{P1}*. If only orders of the three products P1, P2 and P3 are assembled on the assembly line, then there are also only three variable takt times* T_{P1}*,* T_{P2} *and* T_{P3}*. In* Sect. 5.2 *we extend the following calculations to products with option variants (mixed-model assembly line of type 3).*

To calculate the target takt time more realistically, the following formula can be used, it is based on Groover (2014) with own extension:

$$T_j = \frac{A_j}{W \cdot E_{AL} \cdot (1 - (e_{fa} + e_{pa}))}$$

T_j is the takt time or the launching interval for an order j. In short: after T_j, a new order $j + 1$ is placed on the assembly line after order j. A_j is the total assembly workload of order j, W is the number of workers on the assembly line. E_{AL} describes the targeted balancing efficiency (see Sect. 2.6.11); thus, an efficiency value that describes the proportion to which each individual takt can be filled with work content. In operational practice, values greater than 90% have proven to be feasible. We add e_{fa} *and* e_{pa} *to* Groover's (2014) formula, which refer to the part of the factual and personal allowance that must be represented in the takt. This value is usually agreed between the employee representative (i.e., works council) and the company.

If the takt time is specified, the formula can be changed so that the minimum number of workers W can be calculated. Minimum refers to the desired balancing efficiency. The takt time could be specified if a certain target volume is to be achieved with a defined work schedule. In business practice, this is the most common case.

Since with variable takt the launching interval, and thus the distance between orders, is directly determined by the respective variable takt time of orders, this can be determined as follows:

$$D_j = V_{CS} * T_j$$

with

D_j Distance to the next order j or launching distance of order j
V_{CS} Conveyer system speed = speed of the assembly line
T_j Takt time of order j

If it is noted that the variable takt time and the distance D_j *are* always in direct proportion, the speed of the assembly line V_{CS} is defined by the smallest variable takt time T_{min} and the minimum distance of the product in question D_{min}.

$$V_{CS} = \frac{D_{\text{minimum distance between orders}}}{T_{\text{minimum takt}}} = \frac{D_{min}}{T_{min}}$$

4.4.4 Virtual Station Lengths: Effects on the Length of the Stations and the Entire Assembly Line

If products differ by more than 50% in the total assembly workload, this challenge of an assembly line can often no longer be solved in practice exclusively with variable takt, the overlapping areas of the workers become too large (Fig. 4.12), further measures for compensation are necessary.

Figure 4.12 illustrates this effect in a line with three products, assuming that the distances (and thus also takt times) are chosen so that the product with average effort (P_{mean}) corresponds exactly to one station length.

Note on the illustration:

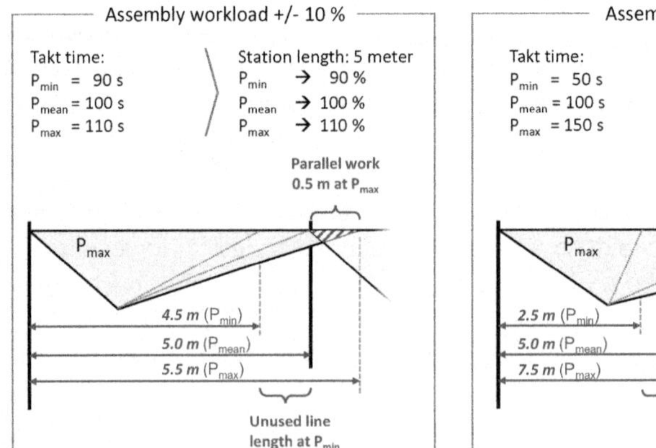

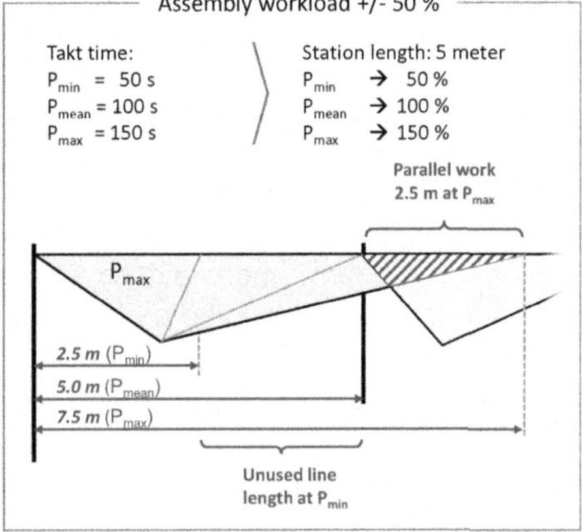

Fig. 4.12 Effects on virtual station lengths and parallel work when spreading the variable takt time

- *A variable takt with a difference of the takt times or distances of +/– 50% (related to the mean workload of P_{mean}) is possible in theory without any problems—the system works with every product-mix.*
- *The greatly reduced distance at P_{min} must be feasible in practice without violating safety regulations of the minimum distances of the products in the assembly line.*
- *In the fixed takt, the number of products and their carriers in the assembly line is constant. In the variable takt, this number fluctuates depending on the product-mix and the level of the takt time difference between P_{max} and P_{min}. The buffer section of the transport system (mostly the return section of the product carriers between the end and start of the assembly line) must be able to compensate for these fluctuations.*
- *If the area of necessary parallel work is greatly increased, in our example 50% of the station length, this is difficult to implement in practice without relevant obstructions between the workers.*

Floor space is one of the most valuable resources for any manufacturing company. For this reason, all measures that lead to an expansion of space requirements must first be critically scrutinized. The variable takt must also face this scrutiny. In the previous chapters, we have always assumed an ideal case. It was possible to design the variable takt in such a way that products with a low work content can be placed closer than one station length and parallel work is possible for products with a high work content. An extension of the stations and thus of the entire assembly line was not necessary. But what if these conditions do not exist? To make it easier to understand, we will explain the relationships using the example of an assembly line with two products and the parameters of station length, speed and distance, which are to be converted from a fixed to a variable takt. The products determining the design of the assembly line are those with the largest and smallest work content (Fig. 4.13).

Initial parameters for the example:
Fixed takt: $T_{Fix} = 500$ s
Variable takt time P1: $T_{P1} = 500$ s
Variable takt time P2: $T_{P2} = 400$ s
Station length: 5 m
Line speed: $V_{CS} = 0.01$ m/s $= 1$ cm/s

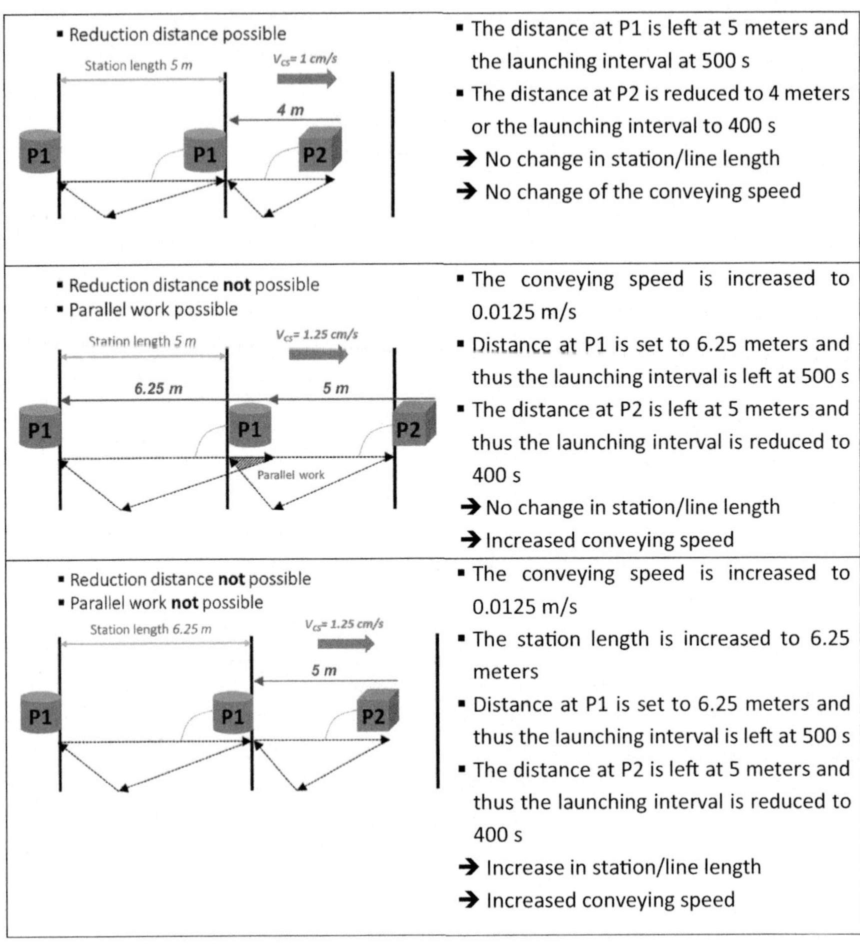

Fig. 4.13 Possible combinations of distances, station length and the conveying speed for realizing the variable takt

From this example, the following implementation steps can be prioritized for a conversion to variable takt:

First of all, an attempt should be made to place the units at shorter intervals so that the station length can remain constant (*priority 1*). If this is not possible or not possible to a sufficient degree to realize the takt or distance difference between the "smallest/shortest" and "largest/longest" product, then parallel work should be used (*priority 2*), i.e., the virtual station boundaries overlap (see Fig. 4.14). If this is also not possible or sufficiently possible, then the stations must be extended (*priority 3*). When combining these three implementation variants, care must be taken to select

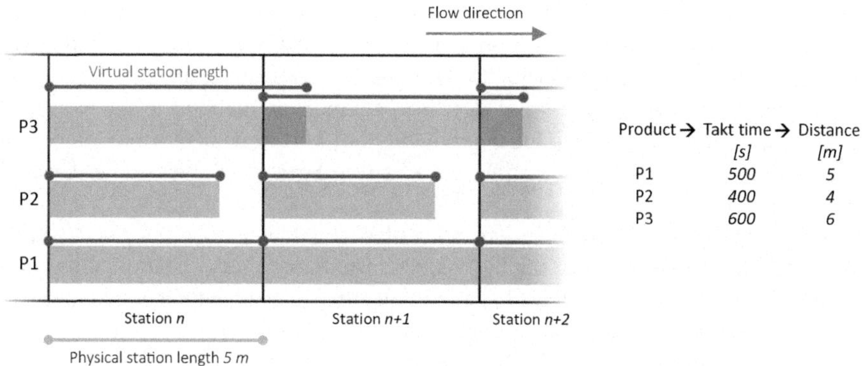

Fig. 4.14 Virtual station boundaries in assembly line design for three products with variable takt

the parameters in such a way that the extension of the station, and thus of the entire assembly line, is minimal.

When using the variable takt, *virtual station lengths* are created due to the different launching intervals or distances depending on the takt time. The physical station boundaries are retained or a decision must be made for a physical station length when designing the assembly line. Or in short: It must be decided after how many meters a station boundary must be drawn on the floor of the assembly hall. The physical structure of the assembly line with its equipment, racks and other facilities is mainly based on these physical boundaries. In Fig. 4.14, we present a line design in which three variable takt times are mapped. The takt time and thus also the distance of the product P1 is oriented to the physical station length of 5 m. A lower takt time of 400 s for product P2 results in a virtual station length of 4 m for product P2. For P3, the virtual station boundaries overlap by 1 m and have a length of 6 m. A new view of variable takt can be derived: **By using variable takt, independent and distinct assembly lines are implemented in one physical assembly line**. In Sect. 4.5 we will discuss this effect in more detail. If no overlaps are desired, it would also be possible in our example to align the physical station length with product P3. In this case, the total length of the assembly line would then increase. Another way to avoid this would be to increase the conveyor speed of the line (see example in Fig. 4.13).

4.4.5 Segmentation of the Product Portfolio into Takt Time Groups

In the current chapter, we use a mixed-model assembly line of type 1 in all examples. This makes it easier for us to explain how variable takt works. In the current era characterized by mass customization (Pine et al., 1993; Kotha, 1995), in which increasingly customized assemblies are required, there are hardly any real examples of a mixed-model assembly of type 1. For this reason, we already combined the variable takt with WATT balancing to the VarioTakt in Sect. 2.5 in order to provide a solution for a mixed-model assembly line of type 3. In addition to the workload differences between the individual products, the workload differences of option variants must now also be taken into account (see Sect. 2.6.8)—customer-specific orders are created.

Ideally, each of these customer-specific orders could be assigned its own variable takt time, this would produce the best results in terms of line utilization. However, each individual order would have to be individually balanced, which would mean an unreasonable administrative effort in planning or in AD for MMA, which is why this option is ruled out in practice for the time being. In addition, it would be impossible for the workers in the assembly line to follow an individual line balance for each order; in extreme cases, a new work sequence would be created for each new order.

To be able to use variable takt sensibly, all orders must be divided into takt time groups. All orders in a takt time group have the same takt time. In the example of Fendt's tractor production in Sect. 4.2, a separate takt time group was selected for each tractor model. In other applications, however, this assignment may not be as clear, purposeful or even feasible. The aim is to find an optimum between the efficiency of the line (= as many individually suitable takt time groups as possible) and the economic effort required for the line balancing of each individual takt time group.

We hope that the content we have described so far has been able to convince other companies or production units to use it or at least to consider using it. Should this be the case, the following questions will arise at an early stage of the planning phase:

- How many takt times or takt time groups are to be selected?
- What is the value of the individual takt times?
- How are the individual customer-specific orders assigned to takt time groups?

Scientific research has not yet made any contributions here known to the authors, and an enumerative solution by means of algorithms had not been described so far. For this reason, we have dedicated a separate Chap. 5 to this issue in our book. In Chap. 5, different approaches to form takt time groups are described; Chap. 5 is based in large on the research work of Moench et al. (2020b).

4.5 Advantages of the Variable Takt

In Sect. 1.2, we derived the current challenges for assemblies from the changing business conditions. These result in the following four target areas that can be positively influenced by the use of variable takt:

1. *Costs in assembly operations*—by reducing utilization losses (Sect. 2.8)
2. *Modification and maintenance costs*—for adjustments to volume, product-mix or design
3. *Investment costs*—due to a reduction of necessary investments during initialization or changes in assembly (e.g., floor space, operating equipment, etc.)
4. To assemble in *high quality*—due to a levelled workload for the workers

We would like to supplement this list with the numerous managerial advantages that arise from continuous flow production, as we describe in Sect. 2.1.

5. *Managerial advantages* due to production in a flow assembly

After more than 4 years of practical implementation of variable takt, we can report numerous benefits in all target areas. On the one hand, we can show that the classic key figures of an assembly, such as productivity or throughput, can be increased despite a high variance in mixed-model assembly of type 3. In total, however, the advantages are even more diverse. Since the assembly line with variable takt adapts independently to its workload, fewer re-balancings are necessary, and in particular when the mix in the product portfolio changes, the variable takt compensates for this. The takt times of individual products or takt time groups can be changed independently of other products, not all products on an assembly line have to be re-balanced at the same time. The output in terms of workload, and in many cases also in terms of revenue, remains more constant despite the changing product mix. Line series-sequence restrictions due to the mix can be significantly reduced, and the sales department will appreciate this. Not to be underestimated are the many managerial advantages (Sect. 2.1) that you get when you succeed in assembling continuously in an assembly line despite the greatest differences in product structure.

4.5.1 Increase in Productivity

Significant productivity increases are possible through the introduction of variable takt, but the size of these increases is highly dependent on the workload differences of the products and on the currently selected line balancing method (maximum takt, flexible worker use, WATT balancing, etc.). A small calculation example: In a mixed-model assembly of type 1, two products are assembled with a distribution of 50% each. The takt is based on product P1, product P2 has a 20% lower workload. Product P2 therefore only has a 20% loss in productivity due to idle time. With a weighting of 50% of the total volume, this results in a 10% loss in productivity for the entire assembly line. With the introduction of the variable takt, this productivity loss is immediately and fully compensated (Fig. 4.15).

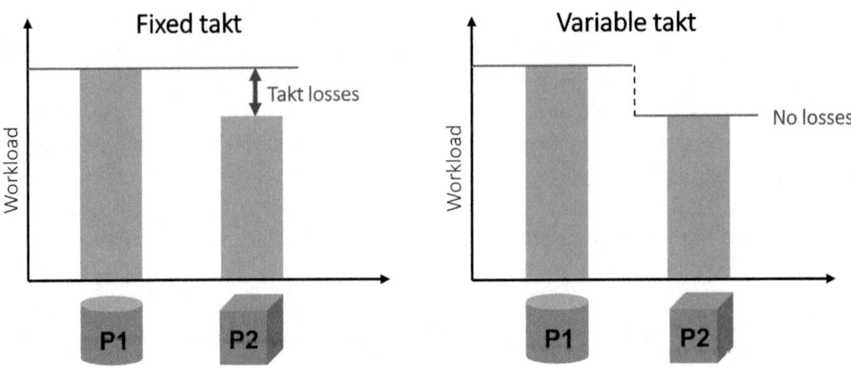

Fig. 4.15 Fixed takt vs. variable takt—elimination of takt losses in mixed-model assembly of type 1

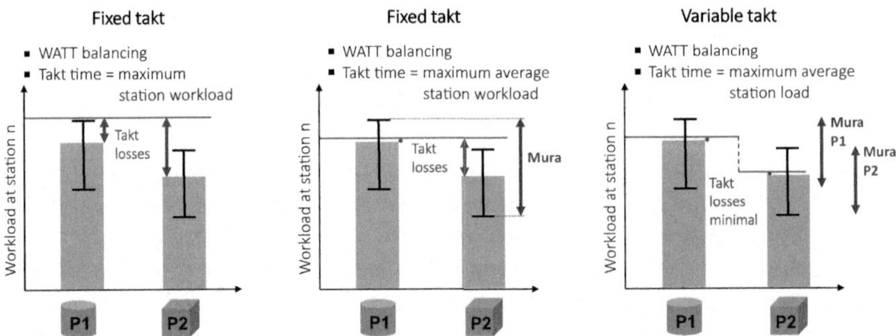

Fig. 4.16 Fixed takt vs. variable takt—elimination of takt losses in mixed-model assembly of type 3

If we transfer these findings to a type 3 mixed-model assembly line and combine the variable takt with the WATT balancing method, the result is the VarioTakt (cf. Sect. 2.5). With the VarioTakt, the takt losses can be completely eliminated with an (1) ideal order sequence (cf. Sect. 3.2.4) and (2) ideal WATT balancing. Ideal means that all WOs can be distributed across all workers in such a way that the average station times correspond exactly to the takt time; in this case, we can speak of a balancing efficiency of 100% (cf. Sect. 2.6.11).

In the case of a classic balanced mixed-model assembly in a fixed takt, which already uses the WATT balancing method, an average productivity increase of 6–10% can be achieved when switching to variable takt. Both takt losses and model-mix losses can be reduced. In Fig. 4.16, we illustrate the reduction in takt losses, and model-mix losses can be further reduced by a lower Mura (see Sect. 3.2.4).

In the scientific literature, Prombanpong et al. (2010) found an increase in productivity of up to 23% for the final inspection process of an automotive final assembly line if it was changed from a fixed to a variable takt. In their 2013 paper, Tong et al. (2013) prove that the classic model-mix losses, such as idle times and overloads, are lower with variable takt than with fixed takt. Liu et al. (2010) also come to a very clear conclusion. In an experiment with three products and 10 workstations, they succeed in proving that variable takt can reduce losses due to idle times and overload by up to 56%. Huchzermeier and Moench (2022) also succeeded in providing a theoretical proof for productivity increases of variable takt in their paper. The cost advantages compared to the fixed takt are proven in further numerous papers (Bard et al., 1992; Fattahi & Salehi, 2009; Moench et al., 2020a).

4.5.2 No Series-Sequence Restrictions: 100% Customer Sequence

Managers of many companies in industrial production are well aware of this situation: one of the major potential conflicts between production and sales are series-sequence restrictions. On the one hand, there is the sales department, which wants to produce every mix constellation with a minimum frozen zone, and this should fluctuate from week to week as much as possible, just as customers demand. On the other side is production, which, due to supply chains, productivity targets and technical bottlenecks, wants to assemble the exact same model-mix throughout the year with as large a frozen zone as possible. The line balancing, and thus the desired balancing efficiency, are designed for a predictable and thus fixed mix. Following our conclusions on mass customization and mass individualization from Sect. 1.2.1, future business success can only be achieved or increased by focusing even more strongly on customer needs. Assemblies must become more flexible and must be able to cover not just one product group but an entire product portfolio. A true build-to-order assembly is becoming increasingly important, but also more complex due to larger and more varied product portfolios in assembly. **With the introduction of variable takt, all sequencing constraints that are rooted in balancing between different products can be eliminated. The series-sequence of the assembly line can be completely oriented to customer demand.**

According to his scientific research and model calculations, Groover (2014) also states that the sequence of the producing units has no significance when using the variable takt. Huchzermeier and Mönch (2022) come to the same conclusion, proving theoretically that any sequence is possible and always under the best conditions of assembly line productivity—ideally: no idle times, no overloading.

4.5.3 Reduction of Re-balancing: Stable Productivity

In a mixed-model assembly with a fixed takt, productivity is very dependent on the model-mix. It is possible to generate very good productivity values for a preset, i.e., forecast product-mix with the WATT method, flexible worker deployment and the use of pre-assemblies. However, if the product-mix changes (seasonal, shifts in demand due to the introduction of new products, recessions, etc.), productivity can change and a re-balancing is usually unavoidable. **By using variable takt, re-balancing can be avoided, as productivity is almost independent of the product-mix on the assembly line.**

The following example calculation in Fig. 4.17 illustrates this effect.
Notes on the sample calculation:

- *The gray fields represent input parameters, all other values are calculated.*
- *The fixed takt does achieve similar productivity values as the variable takt in the model-mix in scenario 5. In practice, however, it will hardly be possible to produce product P1 with a balancing efficiency of 106% and a volume share of 60% without overloading (=line stops) or massive use of flexible workers (=additional unproductivity), model-mix losses will increase significantly.*

The effect of stable productivity independent of the model-mix becomes even more visible when the results of the example calculation are displayed graphically (Fig. 4.18).

							Fixed takt Maximum station workload			Fixed takt WATT balancing			Variable takt		
		P1	P2	P3			P1	P2	P3	P1	P2	P3	P1	P2	P3
Workload		90	70	60			90	70	60	90	70	60	90	70	60
						Takt time	95	95	95	85	85	85	95	75	65
						Balancing efficiency	95%	74%	63%	106%	82%	71%	95%	93%	92%
					Ø			Productivity			Productivity			Productivity	
Model-mix scenario	1	20%	30%	50%	69			73%			81%			93%	
	2	30%	40%	30%	73			77%			86%			93%	
	3	40%	30%	30%	75			79%			88%			94%	
	4	50%	30%	20%	78			82%			92%			94%	
	5	60%	20%	20%	80			84%			94%			94%	

Fig. 4.17 Fixed takt vs. variable takt—change in productivity with changes in model-mix

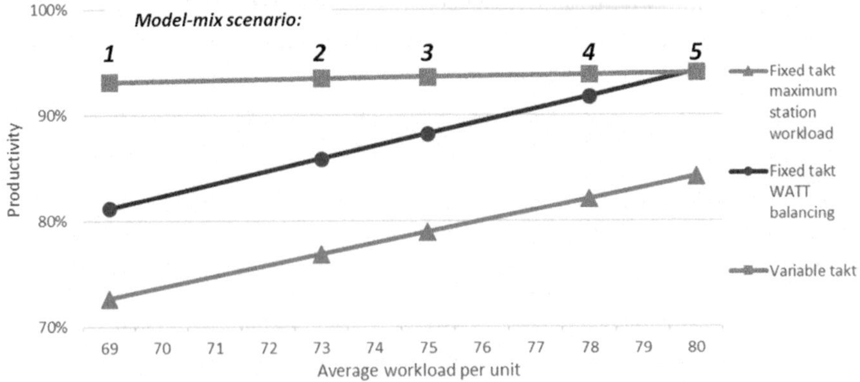

Fig. 4.18 Productivity progression with mix changes based on an example calculation

4.5.4 Reduction of Re-balancing: Constant Output of Workload and Turnover

A mixed-model assembly in variable takt appears unbalanced at first glance (see Sect. 4.3.1), but the workload per time unit is distributed much more evenly. The output of workload is thus more independent of the model-mix. If it is assumed that products with a higher workload generate more revenue (revenue usually even increases disproportionately to the workload), then a model-mix production with the variable takt achieves a more constant output of revenue. Especially for companies with seasonal product ranges, this is of great advantage. In periods when products with a lower workload (lower sales) are more in demand, more units can be produced per unit of time, or the same number of units can be produced in a shorter line operating time. *Seasonal re-balancing to adjust assembly to the changed model-mix is therefore not necessary*—these modification costs can be avoided.

In order to demonstrate this effect on a real example, these effects can be shown for the tractor assembly of the Fendt for a production year. *Note: All of the following data on volume and the corresponding takt times have been changed by a constant factor in order not to publish competition-relevant information. However, this does not affect the significance of the effect of the variable takt.*

Example calculation based on Fendt tractor production

- The Fendt model range is particularly suitable for a variable takt, as the technical scopes of the tractors differ extremely and thus also the total assembly time required. The gross list price of the respective models also has a wide range (basis is the configurator on https://konfigurator.fendt.com as of October 2019).

– Fendt 200 VFP: **95,000 EUR**	– Fendt 700: 240,000 EUR
– Fendt 200 S: 105,000 EUR	– Fendt 800: 265,000 EUR
– Fendt 300: 135,000 EUR	– Fendt 900: 345,000 EUR
– Fendt 500: 170,000 EUR	– Fendt 1000: **385,000 EUR**

- The calculation is based on the data of a real production program, a real assembly line utilization (incl. breakdowns) and revenue values based on the gross list price from the configurator as of the call-off date October 2019.
- For the simulation of the schedule of customer orders with a fixed takt, the fixed takt was selected so that stable production would still be possible. This is the biggest limitation of this example calculation, since this work program has never been executed in a fixed takt. Figure 4.19 shows that the fixed takt was chosen to exceed the smallest type by about 12% and to fall below the largest type by 29%. This already represents a very large stress in the fixed takt. If we were to increase the fixed takt more in our example (make it "easier" to work in it), the advantages would shift even more clearly in the direction of the variable takt in a comparison. Figure 4.20 shows that the average weighted (by number of units per product) variable takt remains very constant in the months February to July, indicating a stable product-mix in these months. From July to the end of the year, a clear drop

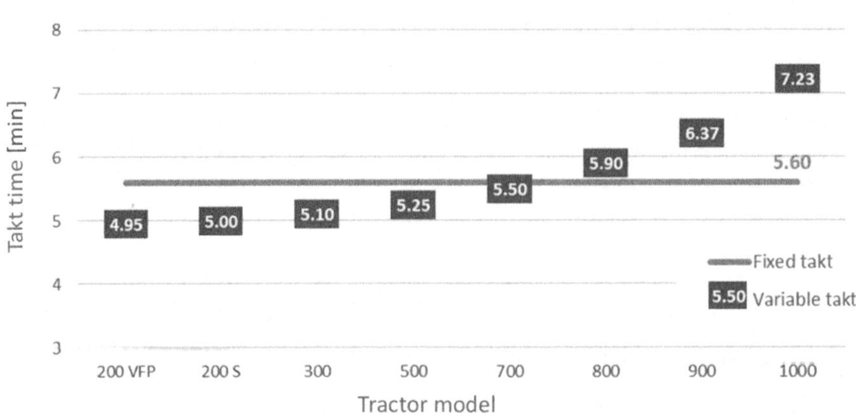

Fig. 4.19 Takt times for the example calculation per Fendt tractor

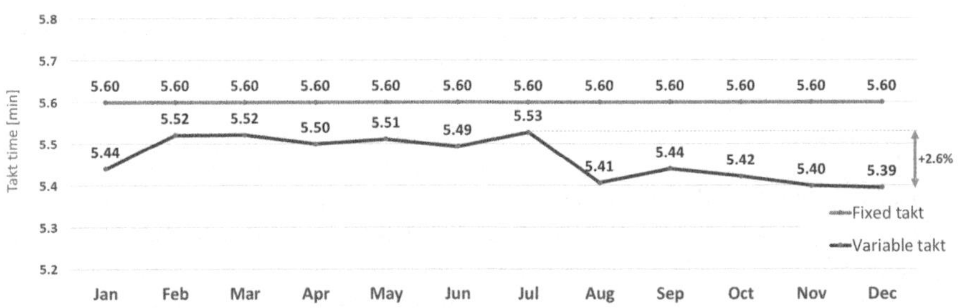

Fig. 4.20 Progression of average takt time per month for Fendt tractor production

can be seen, this is due to a shift in the model-mix towards smaller series in these months. A typical behavior for the tractor market. The variable takt adjusts, the fixed takt remains constant at 5.6 min. Without a shift, the monthly average of the variable takt varies by 2.6% (Fig. 4.20).

Revenue per hour is significantly lower in the months with a "worse" model-mix (more smaller units in December). Revenue per hour also falls off for the variable takt, as the relative difference in list price is significantly greater than between the variable takts. Compared with the fixed takt, however, revenue per hour falls by exactly 2.6% less by which the variable takt adjusts to the model-mix in this period. This is where the variable takt shows its advantage: despite the drop in the model-mix, the variable takt keeps the revenues of the assembly line more constant (Fig. 4.21).

This example calculation is intended to illustrate that when switching to a variable takt, *the total turnover of the assembly line remains more constant even if the model-mix is shifted to lower-turnover units* and thus a costly re-balancing can be avoided or at least delayed.

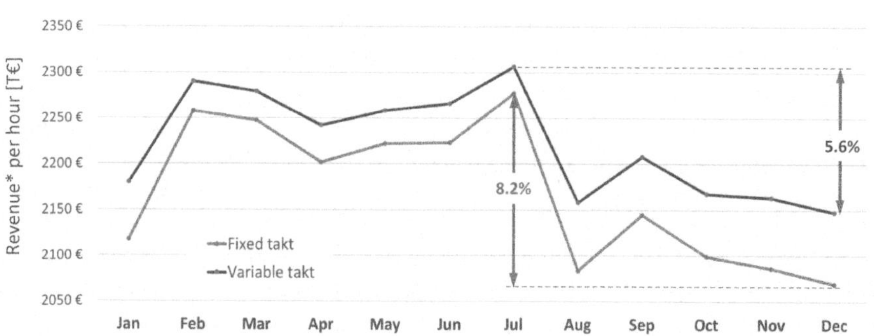

Fig. 4.21 Revenue* per hour of tractor production Fendt (*Revenue based on gross list prices 10.2019)

4.5.5 Levelled Workload for Workers, Minimization of the Use of Flexible Workers, Increase in Quality

Muri and Mura (Sect. 3.1.2) are important parameters in production based on Lean principles. The clearly positive effects of lower Muri and Mura on parameters such as productivity, employee satisfaction, sick leave and, above all, quality are described in numerous literatures (Fujimoto, 1999; Ohno, 2013; Liker, 2021). Experts in assembly design always strive for an assembly design that contains as few fluctuations in workload (Mura) and as few overloads (Muri) as possible. Unlike the fixed takt, *the variable takt reduces imbalances and overload to a minimum.* Due to the use of the WATT in the fixed takt, the use of flexible workers is necessary to compensate for workload peaks. If different products with different total assembly times are assembled on a single line, these utilization peaks increase—Mura increases. When using the VarioTakt, the variable takt compensates for the variance between the products, only the variance of the options in the WATT still has to be compensated for by flexible workers, pre-assembly or drifts. Due to the much more even workload of the workers with variable takt, the use of flexible workers can be minimized, if not completely eliminated. A constant and continuous workload, without overload and underload, of the workers is one of the most important prerequisites for ensuring high quality in the execution of assembly processes (Diwas & Terwiesch, 2009).

4.5.6 Use of Worker Triangle: Lean Management

In the meantime, we should have described in detail that *with variable takt time, the worker, regardless of the model-mix, always starts at the beginning of his/her station with his/her activity at a new unit—there is no drifting*. This is the prerequisite for implementing the principles of a worker triangle (Sect. 2.3.1); material and information can be supplied at the best possible position. Routes remain constant, there is no waste due to drifting workers.

4.5.7 Reducing Investments: Lengthening the Assembly Line Without Actually Lengthening the Assembly Line

In Sect. 4.4.4, we have already discussed the effect of the "virtually extended" assembly line. By deliberately and planned overlapping of the work areas of the employees in variable takt for products with a large work content, *the assembly line is virtually extended—without changing its physical extension*. On the one hand, the "price" paid for this is a lower number of units on the assembly line if there are more products with a large work content on the line, and therefore also a lower number of assembled orders per unit of time. On the other hand, these units transport more value added and revenue, as we have shown in the previous chapters. The reduced output can be compensated for by changing working times. A necessary and very investment-intensive extension of the assembly line can thus be avoided. The following example is intended to serve as an illustration:

A real-life example: A plant in transition from technical product innovations
You are the plant manager of an assembly plant that currently assembles three different products. Your assembly line is balanced with a fixed takt, your products differ in the total work content by +/−5%. You use the WATT method, as well as drifts, flexible workers and pre-assemblies to master this variance. A targeted model-mix in-car sequencing supports the levelling of your assembly. Technical innovation in your industry, which your company translates into new products, gives you a fourth product for your plant. Unfortunately, due to the new technical innovations, the work content of this product is 20% higher than that of your existing products; moreover, the new product will take a significant share of the total production volume. Since you do not want to increase your takt time with the new product in order to keep the output of the assembly line constant, you are forced to increase your assembly line by 20% additional stations and 20% additional workers in order to process the increased content. Not only would this force you to accept significant productivity losses due to an increase in takt losses for your existing products but extending the assembly line is very costly and investment intensive. You would significantly worsen your cost position!

In this situation, if you convert your assembly line to variable takt, you could not only realize productivity gains for your existing products, you could also avoid the necessary extension of the assembly line. You balance your new product similar to product 3 from Sect. 4.4.1, *thus you create overlapping work areas only for this product and extend your assembly line "virtually" by 20%. In addition, you avoid productivity losses for your other products. Because you have correctly understood the principle of variable takt, you are aware that you are buying this "virtual lengthening" with a lower number of orders per unit of time, but this can be compensated for by adjusting the line run rate or working time models.*

In the scientific literature, Bard et al. (1992) describe this positive effect of variable takt and demonstrate it using mathematical models.

4.5.8 Reducing Investments: Horizontal Splitting of the Assembly Line Without Splitting the Assembly Line

In their work, Gans (2008) and Swist (2014), who builds on the findings of Gans, propose a horizontal splitting into two assembly lines, each with an independent takt time, if the processing times (mean values and variance) of two product groups differ significantly. Figure 4.22 shows this solution approach.

The disadvantage of this approach is, in addition to the increased space requirement of two or more assembly lines and the necessary investments, also the lower division of labor due to a necessary increase in takt time when the volume is divided (Swist, 2014) between two assembly lines. In short, if the constant volume is divided between two assembly lines, the takt time in the two separate lines is significantly higher than in one assembly line. Here, variable takt plays out further advantages, because not only two but an (almost) arbitrary number of takt times (or takt time groups) can be generated without having to divide the line horizontally. Thus, a virtual horizontal division into several assembly lines takes place, the disadvantageous takt time increase, due to quantity division in several separate lines, does not occur. Or to put it another way: *By using variable takt, it is possible to map two (or more) different assembly lines with differing takt times and station lengths into one physical assembly line.*

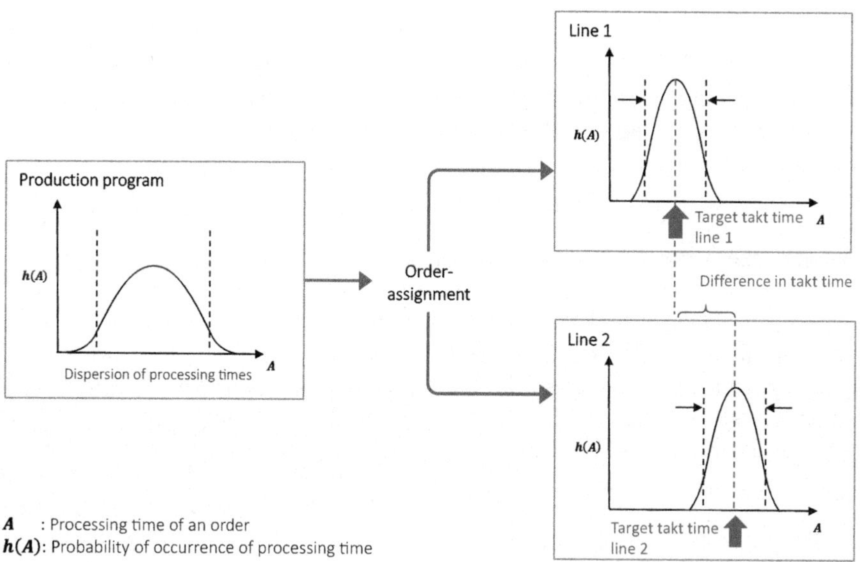

Fig. 4.22 Formation of a multi-line system for variance minimization (adapted from Gans, 2008)

4.5.9 Early Testing of Prototypes and Pilot Series, Targeted Product Launch Curves

In recent years, it has proven very valuable to produce prototypes and pilot series in the future assembly line at the earliest possible stage—despite the currently really great possibilities to virtually simulate an assembly process. Planning errors and bottlenecks are discovered very early on, product adjustments can still be made, improvement ideas from the assembly line employees can still be incorporated into the product design. Unfortunately, very large productivity losses are often associated with this when prototypes are assembled in the production line. The employees on the assembly line are not yet sufficiently trained on the new type; there is a lack of routine in the assembly process. When the prototype is at the respective station, short stops can occur at this station that stop the entire assembly line. These short stops can add up to a large loss over the course of the assembly line run. *With variable takt, it is possible to give exactly this one prototype a "little more" distance and thus more time for each worker.* The loss is thus only present once, and ideally, there are no line stops on the entire assembly line that add up. This methodology could also be used when starting up a new product in a mixed-model assembly. For a limited period of time, the gap, and thus the takt time, is slightly increased for the new type. The accumulation of line stops due to a lack of routine among line workers can be avoided.

4.5.10 Single Re-balancing of One Product Group in the Mixed-Model Assembly Line

Every production manager tries to avoid re-balancing, this represents a large planning and qualification effort and increases the probability of assembly errors. Initially, other measures are used, such as changing working hours or extended shift models. Not only does re-balancing take up capacities and thus costs in the entire organization, but in most cases, re-balancing also represents a peak workload for the entire organization that cannot be handled with the existing capacities of assembly planning and engineering alone. Temporary service providers often fill these gaps. Unfortunately, in fixed takts, all products on the assembly line have to be re-balanced simultaneously when the takt time changes. This is where variable takt comes into its own. Since with variable takt the takt time is a combination of speed V_{CS} and launching distance D_i, by changing the distance of only one product group, the *takt time of exactly this one product group* can be *changed without influencing the takt time of any other product in the mixed-model assembly line*. In practice, this is a huge advantage! This would allow efficiency initiatives for a specific product group in the assembly line to be implemented quickly without impacting other products in the assembly line. Peak loads for the entire organization due to a necessary simultaneous re-balancing of all products could be stretched out and thus avoided.

4.6 Continuous Flow Technologies for Variable Takt

In the previous chapters, it became clear that a technical prerequisite for the implementation of a variable takt is a continuously flowing product stream.

In an intermittent flow assembly (see Sect. 2.2), no variable takt is possible. In these organizational types of flow assembly, work is performed on the stationary product, and at the fixed takt time either all products are moved simultaneously, usually announced via a sound signal, or one after the other by a takt length in the direction of flow. *With variable takt, the employees change the unit to be processed at different times and independently of each other*; this is not possible with intermittent flow operation.

Another prerequisite for the implementation of variable takt is the need to be able to change the distance between the assembly objects to be assembled. This ability to change the distance does not have to exist during the entire flow assembly, because the product-specific distance remains constant during the course of the assembly line (see Sect. 4.2). It is sufficient to be able to set a specific distance when the orders are placed on the assembly line or, in other words, it must be possible to place the products on the assembly line at different intervals.

Automated guided vehicles

AGVs are certainly the most flexible technology for the implementation of variable takt. The technology, which was already developed in the 1950s (at that time still with an internal combustion engine!), was initially used for recurring collective transport for goods over long distances. In recent decades, more and more specialized variants have been developed for a wide variety of logistical tasks. What every AGV system has in common, however, is that the individual units can be controlled independently of one another; it is precisely this capability that makes it so valuable for use in variable takt. Since the description of this technology is not the main focus of this book, we will limit ourselves to comments gained from experience on the introduction of variable takt using AGVs.

The ability to continuously change and adapt an assembly can be a key competitive advantage. On the one hand, when using an AGV for continuous flow assembly, care should be taken to select systems with the least possible mechanical interference with the building structure, e.g., the floor. In this regard, current technologies for navigating the AGV no longer pose a challenge; magnetic loops or optical markings are no longer necessary with the latest systems. The supply of power to the AGV, on the other hand, usually requires the greatest intervention. Inductive charging loops or fast charging points are the current solutions; here, care should be taken to install these only in change-stable areas of the assembly layout, if possible.

The ability to make changes to the control and parameterization of the system quickly and independently of external service providers remains crucial for the successful and adaptable use of an AGV. In this way, even the smallest ideas for improvement from employees and management can be continuously implemented. Often a production consists of a chain of several transport systems, and also several AGVs. Continuous and short-term optimization of control times, run-in distances, control sequences, sequence priorities, etc. is only possible in day-to-day operations by directly assigned and qualified maintenance and technical personnel. The detour via the order processing process of a service provider and the associated internal approval chains slows down and stifles any continuous improvement process.

Overhead conveyor/electric overhead conveyor
A classic area of application for overhead conveyors is automotive assembly, usually in areas where assembly is carried out on the underside of the vehicle. An independent drive unit runs in or on a profile, the data and power supply is provided via slide rails, and the product to be processed is located in a receptacle below the overhead conveyor or on the conveyor. The ability to control the drive units independently of one another is also decisive here for use with variable takt. As with AGVs, the control software should also be able to be independently maintained and modified in order to be able to make quick changes.

Chain conveyor
With the chain conveyor, the production units are continuously transported via a chain, which is usually sunk into the floor. With an intermittently moving chain conveyor, no variable takt can be realized. The product sits on a goods carrier, which in turn can be connected to the chain by various mechanisms. This connection should never be fixed, but always releasable via a mechanism. In a conventional chain conveyor, the chain links, the carriers, are distributed evenly and thus fixed; this distribution corresponds to a takt, station length respectively. The number of carriers thus corresponds to the number of possible stations of the assembly line. However, it is also possible to design the chain in such a way that the number of carriers is maximized. Thus, there are many more carriers in the chain than there are stations in the line. In this way, the distance between the units in the line can be set differently. If a new product is placed on the chain conveyor, the spacing can be changed incrementally. The distances, and thus the takt times, cannot be set as flexibly as desired, as is the case with AGVs, but these incremental distances can also be used to realize a variable takt. The following sketch in Fig. 4.23 illustrates a chain conveyor with products P1, P2 and P3, and the variable takt time distribution in a ratio of 3:2:4. Even with the implementation by means of chain conveyors, there are no series-sequence restrictions, any mix constellation can be represented.

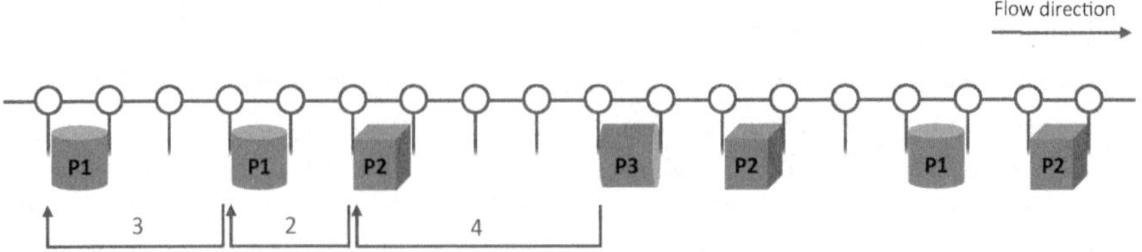

Fig. 4.23 Schematic of a chain conveyor for the realization of variable takt in the example with three takt times

Rope conveyor
The rope conveyor is based on the basic principle of the chain conveyor, except that no fixed carriers are installed in the rope. The carrier can be flexibly attached to the rope, similar to the principle of chair lifts in ski resorts. This technology has not been tested in practice as much as the chain conveyor, but it can be used to solve the limitation of the incremental variable takt of the chain conveyor.

Flat conveyor/conveyor belt
In a conveyor belt, pieces, in our case assembly objects, are continuously moved in a flowing manner on a belt. This conveyor belt can be made of a flexible plastic or rubber, as at the checkout in a supermarket. However, it can also be made of a different material for larger loads, similar to the conveyor line for people at airports. With this technology, a variable takt can also be realized. The products are placed on the conveyor belt directly, or with a carrier, at different intervals.

4.6.1 Apron Conveyor: An Alternative Implementation of Variable Takt Without Distance Adjustments

The apron conveyor, the classic in automotive assembly. The advantages of the apron conveyor are certainly undisputed, which is why it is used in almost all large automotive assemblies. In the apron conveyor, individual large push plates are lined up and moved continuously by means of a chain or friction wheels. The product carriers for the objects to be assembled are located in the center of the push plates. A special form of apron conveyor is presented in Sect. 8.4 with ARC Assembly. It is not possible to implement a variable takt with a classic apron conveyor. If it were possible to insert intermediate plates between the individual push plates, an incremental variable takt, similar to that of a chain conveyor, could be realized. At the time of writing, however, no implementation of this type is known.

An Alternative: The Variable Takt Is Conditionally Realizable in the Apron Conveyor
To implement the variable takt, it is necessary to be able to change the distance between the products on the assembly line. In the following, we will explain a possibility to implement the variable takt also on an assembly line where the distance between the products and their carriers cannot be changed as, for example, on an apron conveyor. The following proposal contains some clear disadvantages and can certainly not be implemented directly in the classic apron conveyor of an automobile manufacturer.

The disadvantages are:

1. Batch production of the product groups is necessary
2. The products must be of sufficient size or cubature to allow increased parallel work on them

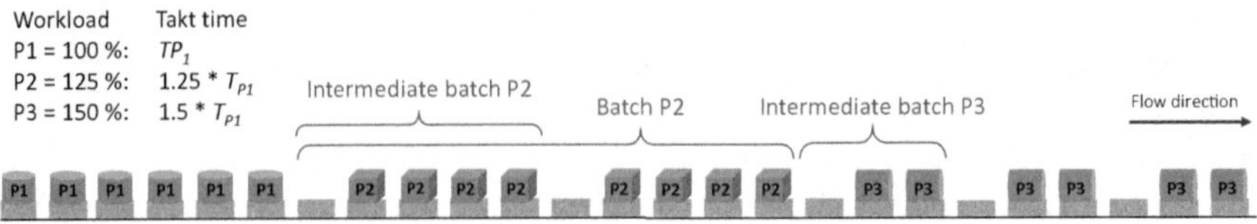

Fig. 4.24 Implementation example of variable takt in the apron conveyer

This solution was successfully used to assemble Fendt tractors and their cabins for over a decade before switching to AGVs.

For the explanation we use an example with three different products on the assembly line (Fig. 4.24).

Relative workload of the products: P1 = 100%; P2 = 125%; P3 = 150%.

The assembly line, station lengths, and takt times are based on the smallest product P1. We call the takt time T_{P1} of product P1 the basic takt time T_1. If we want to realize a takt time of 125% ($T_{P2} = 1.25 * T_1$), then we do not put a unit on the line after every fourth unit—a gap, i.e., an empty unit, is run through the assembly line. 5 stations with 4 units generate 125% of the base takt time per unit. We refer to these four units, including the gap, as the intermediate batch. For P3, it would be two units and one empty unit. Each worker can receive work content in the volume of $R_{w,P2} = 1.25 * R_{w,P1}$ at product P2. For P3, $R_{w,P3} = 1.5 * R_{w,P1}$. Please note again that we are in a batch production in an assembly line. With each takt, the worker drifts 25% of the station length at P2 and 50% of the station length at P3 but makes up this loss completely with the gap.

It can be seen that the different takt time groups cannot be set smoothly, but only incrementally, with the following progression: 200%—150%—133%—125%, and so on. Because the takt times follow the formula:

$$Takt\ time\ T_i = \frac{Number\ of\ products\ P_i\ in\ intermediate\ batch + 1}{Number\ of\ products\ P_i\ in\ intermediate\ batch} * Basic\ takt\ time\ T_1.$$

To realize a takt time of 150%, we thus place an empty unit on the line after two units: 1.5 = (2 + 1)/2 (Fig. 4.24).

4.7 Interlinking of Assembly Sections

In most factories, assembly consists of more than one assembly line. Very often, due to the local conditions of the existing building structure, assembly lines have to be divided. Also, different production technologies like assembly and painting require a division into different assembly sections. There are different approaches to connect these parts of assembly sections; a differentiated management focus on flexibility, productivity, or safety changes the focus of these approaches. If a variable takt is used completely or only in a part of the assembly, additional aspects have to be considered.

If two sections of a line are to be connected to each other, for example, due to a change of building, a decoupling buffer is usually used for this purpose. This decoupling buffer must at least cover the product transport, but in practice, it is usually larger. The reason for this is the desire to decouple disturbances on one section of the assembly line from those on the other section. However, a note to any production manager should be made at this point. As has already been shown in Sect. 2.1, one advantage of continuously flowing processes is that problems and disruptions become visible much faster and more clearly due to a continuous flow. With larger decoupling buffers, this advantage is lost—poor or faulty processes remain undetected for a longer time.

If two assembly lines with variable takt are connected, the products on both lines should always use the same variable takt. Any flexibility in the model-mix can be mapped without any problems. The decoupling buffer can thus be minimized. It is also possible to set the takt times of the individual products differently on the two lines. If the assembly content of the two products is very different on the two assembly lines, this could be an advantage. However, this advantage is bought by a loss of flexibility. Due to the different variable takts of the products on both assembly lines, it must always be ensured that the weighted average of the takt of both lines remains the same. If the weighted average differs, the decoupling buffer runs empty or congestion occurs on the upstream assembly line. With a larger decoupling buffer, the period until these effects occur can be stretched, but it cannot be eliminated.

Another challenge is when an (assembly) section in fixed takt has to be connected with one in variable takt. This could be the case if, for technical reasons, the line section in fixed takt cannot be converted to variable takt. First of all, a decoupling buffer between the two lines is indispensable. The size of the decoupling buffer initially increases the flexibility in the difference of the fixed takt to the weighted average of the variable takt. However, a permanent disbalance still leads to disturbances in the continuous flow operation. If one does not want to accept this loss of flexibility, it is possible to set the takt time of the (assembly) line with lower (assembly) workload somewhat shorter than the weighted average value of the other (assembly) line. This "somewhat faster" naturally represents a loss of productivity and should therefore be kept to a minimum.

Stationary takts

Under certain circumstances, it may not be possible to carry out certain assembly processes in motion, in a continuous flow. These assembly processes must be carried out while stationary. If there is a stationary takt in an assembly line with a fixed takt, it must be integrated into the line by means of the following sequence:

1. At the end of the upstream station, the AGV moves at increased speed to the stationary takt.

2. For the time of the fixed takt (minus the travel times), the unit remains on the stationary position.
3. After the takt time (minus the travel times) has elapsed, the unit returns to the continuously flowing assembly line at increased speed.

This procedure can be implemented in a line with variable takt at the stationary takt, thus generating a fixed takt at the stationary takt. It is also possible to implement variable takt at the stationary takt, but the following effect must be taken into account here: at the stationary takt, the takt time specific distance that is behind the respective order "moves" in front of the order. Products with takt times greater than the average value of the variable takt must move deeper into the next station when leaving the stationary takt in order to reverse this effect. Products with takt times shorter than the mean value of the variable takt must pause at the beginning of the next station after leaving the stationary takt until the distance to the previous unit is restored.

4.8 Experience with the Introduction of the Variable Takt

Conviction

Introducing the variable takt in an organization that has been assembling, thinking, and standardizing in a fixed takt for decades initially requires a lot of persuasive energy and work. The fixed thought patterns of management, executives, planners, and works councils must be severely challenged. There should be a reason for this, a necessity (see Sect. 3.1.1). In addition to good leadership, a recognized need is the best prerequisite for persuading an organization to abandon existing thought patterns. A reason could be the advantages of the variable takt from Sect. 4.5. The following examples could be deliberately brought about in order to support the introduction of the variable takt in a targeted manner:

- Introduction of new product types that significantly worsen the line efficiency of flow assembly by significantly increasing or decreasing the workload.
- The targeted removal of series-sequence restrictions in order to be able to serve market demand in a much more customer-oriented manner.
- Target further, significant productivity increases that can no longer be achieved with classic CIP methods or technical changes.
- The merging of different assembly lines, due to production area reductions, volume decreases, or compensation of seasonal fluctuations.
- Targeted reduction of fluctuating load on employees in the assembly line, with the aim of increasing assembly quality.
- For risk distribution: the construction of distributed local assembly units and thus the combination of different product types with low quantities in one assembly line.

Managers and board members usually listen carefully when it comes to cost savings! Use this feature, calculate and visualize possible savings through improved utilization of the assembly line when using variable takt and VarioTakt. Show the stable assembly costs even in unfavorable model-mix situations.

A prerequisite for convincing all parties involved is the understanding that the variable takt concept is also functional in practice. Here, it is advisable to simulate the new assembly process in variable takt with all those involved. Virtual simulations using software are not necessary for this; they may even be counterproductive, since the managers, planners, and work councils involved will again have to "believe" a software they do not understand. A manual simulation, with visualized assembly lines, magnets, or product models has proven to be very convincing.

The product must be suitable—variance is not equal to variance

The variable takt is the best solution for handling variance in the total volume of work content of different products. Ideally, this should be evenly distributed across the entire assembly line. Option variance that is only reflected at very specific stations on the assembly line is not suitable to be handled with the variable takt. When introducing variable takt, it is therefore important to be aware of this difference so as not to create false expectations in the organization.

Data quality and documented standard work

As is so often the case, data quality is also crucial in line balancing. Thus, all activities should be described in the form of standardized work, ideally on standard worksheets. Digital continuity of this standardized work through to the ERP system would be ideal. In this way, high-quality balancing of the assembly line can be carried out by planners and assembly workers or fully automated by means of software (we give an example of this in Sect. 8.6).

Missing Data: Experience Trumps Software

Today, the line balancing of an assembly line can be supported by increasingly better software solutions, but the most important success factor remains the skills and experience of the planners and workers on the line. A complete precedence graph is required for the fully automated sequencing of an assembly line; creating and keeping this graph up to date still represents a very large effort. Sternatz (2010) investigated the amount of work required to create a digital precedence graph of just one product with 15,000 steps. He calculated a work effort of two man-years! Because of this lack of data, even in the large automotive factories of this world, line balancing is created manually and based on the experiential knowledge of employees, even though there is an increasing trend toward automated line balancing. Despite countless described algorithms for line balancing (Boysen, 2005), these solutions using algorithms cannot be relied upon due to the lack of data. In manual line balancing by workers and planning employees, precedence restrictions are always taken into account, and attempts are made to always locate similar activities of products from other takt time groups at the same station.

Variable volume per day

When using variable takt, the daily number of units produced depends on the model-mix. With the same working time, the number of units produced fluctuates. Although this statement is banal, it must be accepted by management, the organization, and the ERP system. In discussions with managers, the mantra of a constant daily unit output is still too often adhered to. In addition, the existing MES and ERP systems must be able to handle a fluctuating daily unit output.

A new bottleneck: The pre-assemblies

It has already been shown that series-sequence restrictions in flow assembly can almost be eliminated with the variable takt. Unfortunately, this does not apply to pre-assembly and suppliers. Since the decoupling buffers to external suppliers are usually larger, it is only necessary to ensure that their delivery rhythms can adapt to the fluctuations of the variable takt. This can be solved by the buffers and by pre-planning the production program in the variable takt. If pre-assemblies are directly connected to the main line, these pre-assemblies must also operate in the variable takt. If this is not possible, the average takt of the pre-assembly must be smaller than the average variable takt of the main line and decoupled by means of a buffer.

Interlinking and stationary takts

The trickiest challenge in the introduction of the variable takt proved to be the linking of several different assembly lines and production sections, such as AGV lines, chain conveyors in pre-assembly, overhead conveyors in the paint shop. We have already gone into this in detail in Sect. 4.6. The ability of in-house personnel to quickly and easily adjust the programming of various controls of conveyor systems, safety devices, and other equipment on a continuous basis is crucial. In addition, this provides significantly more understanding and know-how about the company's own equipment and software.

4.9 Why Does the Fixed Takt Dominate? Observations and Hypotheses

Given the many advantages of variable takt, the question arises as to why almost all flow assemblies still use a fixed takt. If the methodology of variable takt is so clearly superior to fixed takt in many respects (Sect. 4.5) and research papers, albeit few, have been published for over 40 years (Dar-El, 1978): why does (almost) no company apply variable takt? From the feedback of many presentations and discussions on VarioTakt to production experts, but also researchers in the field of flow production, we can hypothesize the following:

It was not necessary until now!

In the last decades of production, in automotive production as a pioneer in particular, the focus has been on economies of scale through the largest possible (volume) and uniform (same products) production. Initially via single-product flow assembly and with increasing mass customization model-mix assembly lines of type 2 and now of type 3 (Sect. 2.2) will rise in use. Every effort was made to avoid variance in and between products or to assemble them separately.

Three current developments are pushing flow assembly in a fixed takt to its limits with their different concepts for mastering variance.

First, significantly greater structural differences in products due to increasing technology diversification, such as autonomous driving (many additional control units and sensors) or new drive technology (hybrid, electric, etc.) needs to be installed.

Second, an increasing targeted more global distribution of production to manage risk, thus a trend towards smaller standardizing assemblies for several more diverse products has started to develop (Sect. 1.2.2).

And third, a clear drive toward mass individualization (Koren, 2021) and thus toward the assembly of even more individualized products (Sect. 1.2.1) is here to stay.

All these developments lead to smaller, globally distributed, and locally organized assemblies, which increasingly have to represent the most diverse portfolios of products in an assembly line.

Management does not get it!

In recent years, we have explained the VarioTakt to many companies, managers, and planners using Fendt tractor production as an example. With a successful practical implementation behind us, we succeeded in convincing all visitors to Fendt production of the feasibility and advantages of variable takt and VarioTakt. It proved to be much more difficult to explain the principles in purely theoretical terms; lack of understanding was often the result. When the "enlightened" people returned to their companies, they often failed to convince their management. The mantra of the fixed takt is still too deep-rooted. "*That doesn't work!*" "*That creates traffic jams,*" "*Will the AGVs overtake each other?*" "*That's not in line with our Lean principles,*" "*But we need a fixed output,*" to name just a few of the prejudices. This book is our contribution to help eliminate these prejudices and to work on spreading the idea of variable takt.

The interlinking of different production sections becomes more complex!

In Sect. 4.7 we have already described the different possibilities to connect different assembly systems and assembly lines with each other. When using a fixed takt, this connection is a transfer or sorting buffer. The size of this buffer depends on two criteria. On the one hand, it is used to represent a desired temporal decoupling of the two sections, so that line stops or different pause times can be compensated. On the other hand, it is used as a sequencing buffer if a different series-sequence is required in the different assembly sections.

When variable takt is used, a further criterion is added. If the production sections are connected with different variable takt times, it must be ensured that this connection does not run empty or overflow, even over a longer period of time. Losses due to idle times in the upstream or downstream area should be avoided. The different output rate of the two lines is not synchronous if different variable takt times are used for both assembly lines. The output rate changes with the product mix. If different variable takt times are used between production sections, the interlinking of these sections becomes much more complex.

It was not yet technically possible!

A prerequisite in the implementation of the variable takt, is the ability to change the distances of the production objects in the flow assembly. Many existing assembly lines such as chain or apron conveyors do not have this capability, it was not necessary until now. Variable takt is thus not possible. When a product is phased out, the technical equipment is usually also used for the successor product, so there is no possibility of introducing a variable takt for new products. With the increasing spread of driverless transport vehicles or other more flexible transport systems (e.g., electric monorail systems), the technical prerequisites for a variable takt are being created more and more.

ERP and MES systems cannot handle variable takt and variable target quantity!

In many cases, we in production have become slaves to our ERP and MES systems. These systems are not configured to fit our use case, but existing processes have to be adapted to these systems. Actually, it should be the other way around, right? Many existing ERP, MES, or even line balancing systems cannot handle a variable takt and associated variable daily production output. It is not provided for, everything is based on a fixed takt and associated constant output. In the age of increasing digitalization and customer individualization, this way of thinking will quickly prove to be outdated.

Conclusion on variable takt

In this chapter, we now revealed the secret of how variable takt works—it is easy, even simpler as thought, isn't it? However, as always and in every organization, it is not the idea that counts, but the will and the ability to implement it!

For this reason, we deal with further implementation examples in the following second half of our book. We present concepts for the structured introduction of variable takt and combine variable takt with current applications from industry for mastering the variance in assemblies.

References

Bard, J. F., Dar-El, E., & Shtub, A. (1992). An analytic framework for sequencing mixed model assembly lines. *International Journal of Production Research, 30*(1), 35–48. https://doi.org/10.1080/00207549208942876

Boysen, N. (2005). *Variantenfließfertigung* (294 p.). Springer. https://www.springer.com/gp/book/9783835000582

Boysen, N., Fliedner, M., & Scholl, A. (2007). A classification of assembly line balancing problems. *European Journal of Operational Research, 183*(2), 674–693. https://doi.org/10.1016/j.ejor.2006.10.010

Boysen, N., Fliedner, M., & Scholl, A. (2009a). Assembly line balancing: Joint precedence graphs under high product variety. *IIE Transactions, 41*(3), 183–193. https://doi.org/10.1080/07408170801965082

Boysen, N., Fliedner, M., & Scholl, A. (2009b). Sequencing mixed-model assembly lines: Survey, classification and model critique. *European Journal of Operational Research, 192*(2), 349–373. https://doi.org/10.1016/j.ejor.2007.09.013

Dar-El, E. M. (1978). Mixed-model assembly line sequencing problems. *Omega, 6*(4), 313–323. https://doi.org/10.1016/0305-0483(78)90004-X

Diwas, S. K., & Terwiesch, C. (2009). Impact of workload on service time and patient safety: An econometric analysis of hospital operations. *Management Science, 55*(9), 1486–1498. https://doi.org/10.1287/mnsc.1090.1037

Fattahi, P., & Salehi, M. (2009). Sequencing the mixed model assembly line to minimize the total utility and idle costs with variable launching interval. *The International Journal of Advanced Manufacturing Technology, 45*. https://doi.org/10.1007/s00170-009-2020-0

Fujimoto, T. (1999). *Evolution of manufacturing systems at Toyota* (400 p.). Taylor & Francis.

Gans, J. E. (2008). *Neu- und Anpassungsplanung der Struktur von getakteten Fließproduktionssystemen für variantenreiche Serienprodukte in der Montage* (139 p.). University of Paderborn.

Groover, M. P. (2014). *Automation, production systems, and computer-integrated manufacturing* (816 p.). Pearson Higher Education.

Huchzermeier, A., & Moench, T. (2019). *Reexamine our suspense on variation: How variable takt times solve the productivity puzzle* (4 p.). Unpublished working paper. WHU - Otto Beisheim School of Mangement.

Huchzermeier, A., & Moench, T. (2022). Mixed-model assembly lines with variable takt and open stations: The ideal and general cases. *Unpublished working paper. WHU – Otto Beisheim School of Management, 20*(3), 35 p.

Koren, Y. (2021). The local factory of the future for producing individualized products. *The BRIDGE, 51*(1), 100 p. https://www.nae.edu/251191/The-Local-Factory-of-the-Future-for-Producing-Individualized-Products

Kotha, S. (1995). Mass customization: Implementing the emerging paradigm for competitive advantage. *Strategic Management Journal, 16*, 21–42. https://doi.org/10.1002/smj.4250160916

Liker, J. K. (2021). *The Toyota way* (449 p., 2nd ed.). McGraw Hill.

Liu, R., Lou, P., Tang, D., & Yang, L. (2010). A hybrid immune algorithm for sequencing the mixed-model assembly line with variable launching intervals. *Information Computing and Applications*, 399–406. https://doi.org/10.1007/978-3-642-16336-4_53

Moench, T., Huchzermeier, A., & Bebersdorf, P. (2020a). Variable Takt times in mixed-model assembly line balancing with random customization. *International Journal of Production Research*. https://doi.org/10.1080/00207543.2020.1769874

Moench, T., Huchzermeier, A., & Bebersdorf, P. (2020b). Variable takt time groups and workload equilibrium. *International Journal of Production Research*. https://doi.org/10.1080/00207543.2020.1864836

Ohno, T. (2013). *The Toyota production system* (176 p.). Campus Verlag.

Pine, B. J., Bart, V., & Boynton, A. C. (1993). Making mass customization work. *Harvard Business Review Online*. Accessed April 29, 2021, from https://hbr.org/1993/09/making-mass-customization-work

Prombanpong, S., Dumkum, C., Laoporn, S., & Satranondha, E. (2010). A mathematical approach to design launching pattern for a mixed-model production line. *Asia Pacific Regional Meeting of International Foundation for Production Research*, 1160–1164. https://doi.org/10.4028/www.scientific.net/AMR.875-877.1160

Sternatz, J. (2010). Automatic clocking without creating a precedence graph. *MTM User Conference*, 04/16/2010.

Swist, M. (2014). *Taktverlustprävention in der integrierten Produkt- und Prozessplanung* (328 p.). Apprimus Verlag.

Thomopoulos, N. T. (1967). Line balancing-sequencing for mixed-model assembly. *Management Science, 14*(2), B-59-B-75. https://doi.org/10.1287/mnsc.14.2.B59

Tong, K., Xu, K., & Zheng, Y. (2013). Sequencing mixed-model flexible assembly lines with variable launching intervals. *Journal of Shanghai Jiaotong University, 18*, 460–467. https://doi.org/10.1007/s12204-013-1420-3

Womack, J. P., Jones, D. T., & Roos, D. (1990). *The machine that changed the world* (323 p.). Westview Press.

5 Takt Time Groups in the Variable Takt

In the descriptions of how variable takt works in Chap. 4, we have not yet answered a crucial question: How are the takt time groups and the associated takt times of the relevant assembly portfolio determined? According to our previous analyses, implementations, and experiences with variable takt, this is one of the most complex questions in this context. For this reason, we have dedicated a separate chapter to this issue. We start with a presentation of visualizations of takt time groups at Audi. We extend this with the help of our understanding of products and options from Sect. 2.4 and derive a first qualitative procedure for the determination of takt time groups. In Sect. 5.3, we introduce the workload equilibrium view, derive a model for defining takt time groups based on it, and create a heuristic algorithm—the VTGA. These elaborations are mainly based on the work of Moench et al. (2020). We summarize the findings of all sections of this chapter with the VTGTP (*V*ariable *T*akt time *G*roup *T*ransformation *P*rocess), which we use to describe an approach to determine the essential parameters when converting an assembly from fixed to variable takt. Finally, we apply the VTGTP in a case study of the introduction of variable takt in the assembly of Canyon Bicycles and show the great potential of variable takt to increase productivity.

5.1 First Approaches: Visualization of Takt Time Groups

When the variable takt time was introduced at Fendt, the question of an analysis method for dividing the product portfolio into takt time groups did not yet arise. Since each product, in this case each tractor model (Sect. 4.1.1), was documented and balanced in a separate SWS (standard worksheet), the structure of the tractor models was adopted as the structure of the variable takt time groups. Each tractor model was given its own takt time. The clear structuring according to products and options (Sect. 2.4.1) and the realization of the relevance of the segmentation into takt time groups emerged only later. Many very good theories emerge from practice and from the need for initial applications.

P. Bebersdorf, A. Huchzermeier, *Variable Takt Principle*, Management for Professionals, https://doi.org/10.1007/978-3-030-87170-3_5

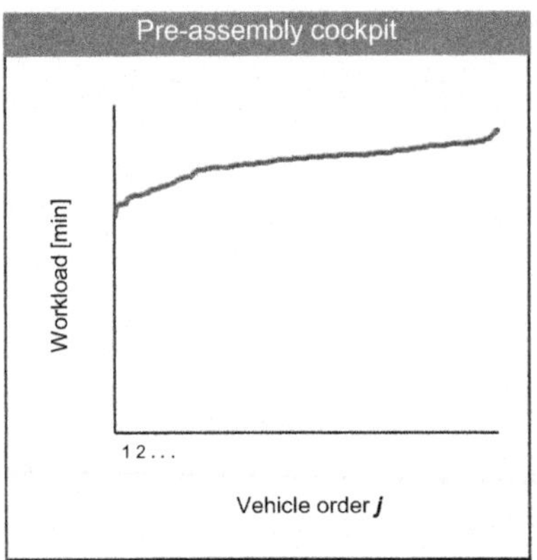

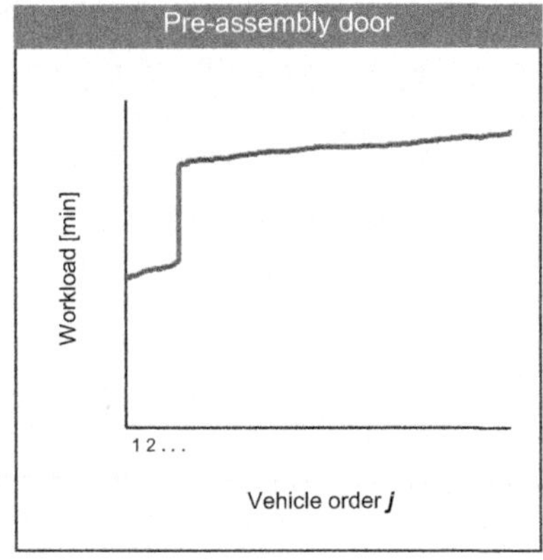

Fig. 5.1 Varying assembly times within pre-assembly after application of the sorting rule (adapted from Hiller, 2018)

Inspired by a technical exchange with Fendt, the first feasibility studies were carried out at Audi on the use of the variable takt. The need for a sensible segmentation into takt time groups was quickly recognized, and it was necessary to develop a generalized method that could be used in as many applications as possible. An initial method of analysis was to record the total assembly workload of each individual order, sort it in ascending sequence, and display it graphically (Hiller, 2018; Fig. 5.1). The number of selected customer orders should be between a daily and a weekly production; a higher number does not lead to a more substantial gain in knowledge. It is crucial that the distribution of products and options in the data set under consideration is representative of the intended planning period of the assembly line. The scope of the observations should be both the assembly as a whole and divided into individual assembly sections. With this type of visualization, a very good initial overview of the distribution of the assembly workload of the individual order variants in the entire assembly and in the respective assembly sections is generated.

We use the index i, $i = 1 \ldots I$ as index of the takt time groups. If $I = 1$, then only one takt time group exists, which corresponds to a fixed takt. We use the index j, $j = 1 \ldots J$ *as an index* of the individual customer orders in the period under consideration.

In Fig. 5.1, Hiller (2018) depicts two assembly sections of Audi at the Ingolstadt production site in Germany. These are two pre-assembly lines of the main assembly of the site with three vehicle models: vehicle A as a 4-door and vehicle B as a 4-door and a 2-door. During the period under consideration, the assembly times of all orders for the respective assembly section were calculated and then sorted in ascending sequence (Fig. 5.1); this is a data set with several hundred orders. It is easy to see that the curve for the assembly workload A_j of cockpit pre-assembly is basically very homogeneous. A steady increase can be seen for the first 200 units, after which a plateau is almost reached. At the end of the curve, a few somewhat more complex cockpit variants can be identified. A slight S-shape can be seen in the curve, a pattern that is reflected in many data sets, as can also be found in the graphs from Fendt in Sect. 3.2.5. On the basis of this representation, it cannot initially be decided whether a meaningful segmentation into takt time groups is possible. This is more clearly visible in the graphic on the right side of Fig. 5.1, where a clear discontinuity can be seen after the first units. This discontinuity is due to the technical difference between the coupe variant of vehicle B (2-door) and the sedan variant of vehicles A and B (4-door). The use of variable takt times could achieve positive effects. According to Hiller (2018), the takt times of the two takt time groups can be defined as follows:

Balancing according to maximum variant:

$$T_i = \frac{Largest\ assembly\ workload\ of\ takt\ time\ group\ i}{Number\ of\ workers} = \frac{A_{i\_max}}{W}$$

Balancing according to median:

$$T_i = \frac{Median\ assembly\ workload\ of\ takt\ time\ group\ i}{Number\ of\ \text{workers}} = \frac{A_{i\_median}}{W}$$

T_i Takt time of takt time group i

Limitations of the method

(a) This procedure assumes an ideal line balancing with a line efficiency of 100%. This is almost never achievable in practice, so the target takt must be set higher for implementation. In practice, a balancing efficiency of 95% has proven to be challenging, but feasible; we have already indicated this in our formula for calculating the takt time in Sect. 4.4.3.

(b) On the one hand, existing assemblies or assembly lines already have their own identity. If the assembly line has to handle many variants and has already been optimized for a fixed takt over a longer period of time, it can be assumed that the full potential of the variable takt cannot be read from the available data. Certainly, assembly contents of complex variants have already been shifted to pre-assemblies in order to minimize takt and model-mix losses in the fixed takt. In most cases, these contents cannot be processed as efficiently in the pre-assemblies as in the assembly line. In Sect. 2.1, we described the advantages of processing work content in the flowing, clocked assembly line.

(c) On the other hand, the work contents of the products with lower workload were probably already enriched with pre-assembly activities in order to fill the takt and achieve the lowest possible takt compensation. With variable takt, this content could be shifted back to pre-assembly, the takt time of these products could be shortened, and thus more units could be produced in the same runtime of the assembly line. This potential is not visible from the current data.

(d) If the takt time T_i of the takt time group i is determined according to the maximum variant A_{i_max} of the takt time group, large takt losses occur.

(e) If the takt time T_i of the takt time group i is determined according to the median value A_{i_median} of the takt time group, the takt losses are significantly reduced. However, we have already described the necessity of limiting the maximum overtakt in the WATT method in Sect. 2.6.9. This is not considered in the present approach. Since no target values of a balancing are given, it can also not be assumed that the suitable takt time is found with the median value.

Conclusion Based on this first pragmatic analysis method, the order portfolio can be divided into takt time groups if two (or more) jumps or discontinuities in the sorted data are clearly visible. This is a highly simplified, but first step towards the definition of takt time groups. If no clear structure is visible and explainable in the visualization, then the analysis method must be extended.

5.2 Qualitative Formation of Takt Time Groups by Orientation to Product Groups

The following example illustrates a qualitative segmentation into takt time groups. Let us take an assembly line with $J = 78$ orders in model-mix of type 3 with $W = 100$ workers on which three different products P1, P2, and P3 with different option variants are to be assembled. The spread between the order with the smallest and largest assembly workload is approximately 30%. If this assembly line is to be transformed from a fixed takt to a variable takt, the first question is the number of takt time groups and the respective takt times. The simple sorting rule is applied via the number of orders J of all product groups, the orders are then qualitatively divided into two takt time groups. This could result in a distribution like the one shown in Fig. 5.2.

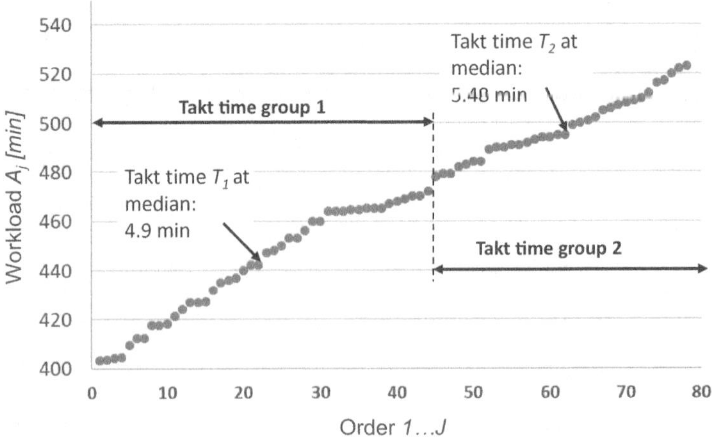

Fig. 5.2 Example—qualitative segmentation into two takt time groups according to sorting of assembly workload A_j

The takt times are formed on the basis of the median value of the respective group. For the formation of the takt time, we use the formula from Sect. 4.4.3 and so $T_i = A_{i\_median}/W$. In order to determine more realistic takt times for practical applications, the extended formula from Sect. 4.4.3 should be used, which takes the desired balancing efficiency and allowance times into account.

In our example, this results in the takt time T_1 with the input parameters:

$A_{1\_median} = 442$ min: Assembly effort of the median value in takt time group 1
$W = 100$; Number of workers
$E_{AL} = 0.95$; Desired balancing efficiency
$e_{fa} = 0.03$; Factual allowance time
$e_{pa} = 0.02$; Personal allowance time

$$T_1 = \frac{A_{1\_median}}{W \cdot E_{AL} \cdot \left(1 - \left(e_{pa} + e_{fa}\right)\right)} = \frac{442 \text{ min}}{100 \cdot 0.95 \cdot (1 - 0.05)} = 4.9 \text{ min}$$

Why is the median and not the arithmetic mean used? Outliers or extreme values in the workload of individual orders should not increase the takt time for all other orders. In the course of our investigations, it became apparent that in existing assembly lines there are always individual orders that show extreme values in the workload. These are generated by workload-intensive rare option variants. The workload peaks should not be intercepted with the variable takt, it would increase the takt time of the entire takt time group, lower the balancing efficiency and still never be sufficient to level the workload peak. When the individual SWS are balanced (Sect. 2.6), these workload peaks must be handled with other measures, such as shifting to pre-assembly (Sect. 7.1.2) or the use of special task workers (Sect. 7.1.3). In Sect. 7.1, we show a practical application with the "Fendt assembly system," which is organized exactly according to this principle.

Dividing the products into two takt time groups with a takt time $T_1 = 4.9$ min and $T_2 = 5.48$ min would be a possibility at first glance, but could lead to significant distortions in practical implementation. The total time of each order consists, on the one hand, of the basic time of its product P1, P2, or P3 and, on the other hand, of the customer-dependent option variants. Thus, the total assembly times of orders from different product groups can overlap—which in practice they do in most cases. In our example, if the entire production program is divided into two takt time groups, as in Fig. 5.2, this can result in orders for product P2 being in both takt time groups—depending on the option variant. Product P2 (or even P1 and P3) would have to be balanced twice, this represents a double administrative effort. For our three products, there would therefore be 6 line balancing variants and thus 6 SWS for each worker on the assembly line (see Sect. 2.7). Each worker would have two different work sequences for each product. This must be avoided. In addition, it will lead to the same material for a product having to be supplied and balanced at different stations on the assembly line.

A qualitatively better decision is made when sorting by products or product groups before sorting by the total assembly workload. Decision criteria to divide the entire portfolio of an assembly line into products have been described in detail in

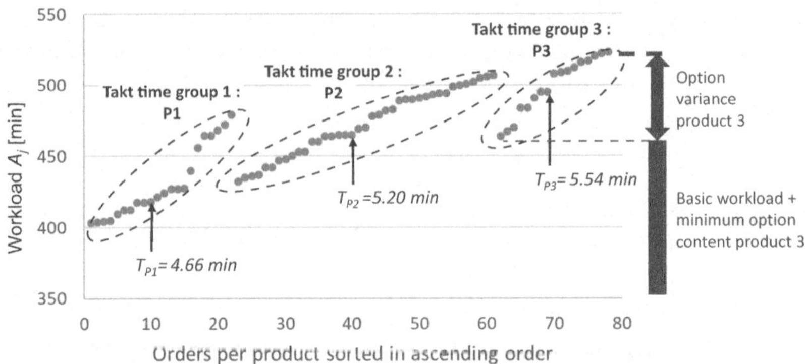

Fig. 5.3 Example—segmentation into $I = 3$ takt time groups according to sorting rule: (1.) product group and (2.) assembly workload A_j

Sect. 2.4.1. If we use the same data from Fig. 5.2 and assign each product to exactly one takt time group, the Fig. 5.3 results from the changed sorting.

Note 1 *In our example, we have decided to divide into three takt time groups—each product has its own takt time group. It would also be possible to combine two products in one takt time group. For example, P2 and P3 could be combined in one takt time group.*

Note 2 *In the groups of products, the spread of the assembly workload due to different option variants can be clearly seen. In order to master these option variants, we combine the WATT method with the variable takt to form the VarioTakt. In a real implementation this is usually not sufficient, variants with a low probability of occurrence* P_l *(see* Sect. 2.6.8*) lead to poor balancing efficiencies, large local workload peaks cannot be levelled even with the WATT. Further elements from the AD for MMA such as pre-assembly, special task workers, etc. are necessary. We describe a concrete practical implementation in* Sect. 7.1.

A segmentation into three takt time groups with takt times $T_{P1} = 4.66$ min, $T_{P2} = 5.20$ min and $T_{P3} = 5.54$ min is now the better solution. In this context, better is understood as a tradeoff between the expected efficiency gain with variable line balancing and the administrative effort for creating and maintaining the line balancing. In our example, there would be three line balances—one line balance for each product and thus one SWS per product and station for each worker. This method can be used for the entire assembly line, but also for feeder pre-assembly lines. If the main assembly line consists of several segments, these segments should be examined individually. If different takt time groups are used for these assembly segments and pre-assemblies, they must be decoupled with buffers. This can lead to a loss of flexibility in the model-mix, a more detailed consideration of this issue can be found in Sect. 4.7.

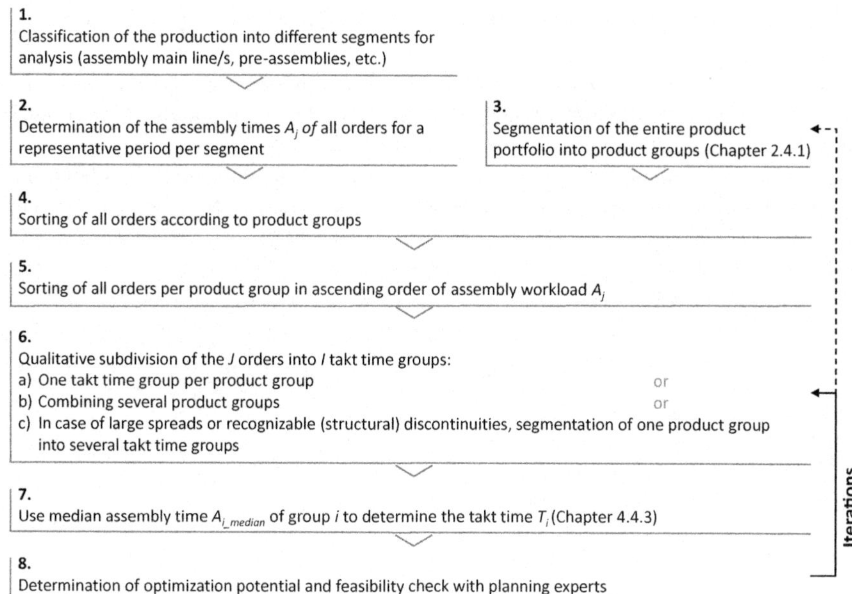

Fig. 5.4 Qualitative procedure for determining takt time groups with extension of presorting by product groups (adapted from Hiller, 2018)

The result of this method should always be discussed with planning experts of the respective area (Hiller, 2018). They can estimate the relationship between the optimization potential through more takt time groups and increasing complexity in administrative effort very well. It can also be discussed with the planning experts whether the selected takt time groups (number and intersections) are feasible for the workers on the line (clearly distinguishable and controllable with regard to qualification). In summary, the procedure can be structured as in Fig. 5.4.

Limitations of the method

(a) As described in the limitations of the method from Sect. 5.1, when using data, an assembly line optimized for the fixed takt, the full potential of the variable takt cannot be extracted. (For a detailed description see the same section in Sect. 5.1.)

(b) As described in Sect. 2.6.9, the maximum overtakting of the WATT should be limited. Very often, this is also a requirement of employee representatives. If the takt time T_i of the takt time group i is determined according to the median value of the assembly time A_{i_median} of the takt time group, overtakting of individual customer-specific orders cannot be limited.

5.3 Variable Takt Time Groups Through Work Equilibrium

While the previous sections described qualitative methods for forming takt time groups, we present a quantitative method in the following. This chapter is essentially based on the publication by Moench et al. (2020).

To ensure consistency of understanding, we continue to use the same indices and build on the findings and definitions of the previous sections. We use the following simplifications to explain the method (Moench et al., 2020):

- One worker w per station n; $N=W$.
- There are J customized orders with an individual assembly effort of A_j per order j.
- All activities are perfectly balanced over all stations. The individual station time a_j *is* thus obtained by dividing the total assembly time A_j by the number of stations, $a_j = A_j/N$.
- The speed of the assembly line V_{CS} is constant at all times.
- No buffers between the stations.
- No setup times between the orders or if available, then these are included in A_j.
- No change of sequence in the line, first-in-first-out principle.
- Open station boundaries.

Under these assumptions, we introduce the weighted average workload per station (Moench et al., 2020) and denote it as *watt* (weighted average tact time):

$$watt = \frac{\sum_{j=1}^{J} A_j}{N * J}$$

If the takt time $T = watt$ is set, the workers drift in the flow direction of the assembly line for all orders with $a_j > watt$. There is an overload situation (Muri)—see Sect. 3.1.2. For orders with $a_j < watt$, there is an underload situation, the workers drift off against the flow direction.

5.3.1 The Work Equilibrium

In Chap. 3 we explained the basic idea of Heijunka. In order to achieve the highest possible productivity of the assembly line, according to or through Heijunka, a balance between overload and underload should be aimed for. The *watt* follows this principle, because the amount of overload, orders above the *watt*, is equal to the amount of underload, orders below the *watt*.

The workload equilibrium does not protect an assembly line from excessive drifting of workers in every sequence, so that even with open station boundaries the final station boundary (Sect. 2.6.10) can be reached and a line stop occurs. The factor β describes the maximum allowed overtakting per order and thus the maximum drift per unit in the flow direction. If methods of mixed-model sequencing

(Sect. 3.2.3) should be applied to minimize model-mix losses, the limitation of overtakting by means of β facilitates these planning efforts. Very often, β is also a parameter for the balancing of the assembly line, which has to be agreed with the workers' representatives.

To illustrate the effects of β, Moench et al. (2020) use real assembly data as an example to show the effects of changing β. Figure 5.5 shows the maximum drift range Υ with different β when $T = watt$.

What happens to the work content above the limit of $watt^*(1 + \beta)$?

Variant 1

This work could be transferred to pre-assembly or to external suppliers. Another possibility would be to use special task workers (see Sect. 7.1.3), i.e., specialists who cover these option variant activities independently of the fixed station employees in the assembly line. Or to put it another way: this is a kind of a flowing pre-assembly in the assembly line. Whichever method is chosen, the aim is not to assign the assembly content to the workers on the line.

Figure 5.6 shows such a line situation. These are the real assembly times of a weekly production of tractor assembly. Due to the workload equilibrium in the use of the *watt*, the overload (area C) is equal to the underload (area B). From this condition, the lower bound $watt^*(1\text{-}\alpha)$ is obtained. The takt time T is set equal to the *watt*. Area D represents the activities that are not allowed to be mapped in the assembly line due to the limitation by β. Following the logic of the *watt*, the areas A and D are equal. If now the activity shares of area D are taken from the orders (the A_j of these orders are reduced), the *watt* is also reduced. Thereupon, new activity components enter the area D. If this procedure is continued, the boundary $watt^*(1\text{-}\alpha)$ moves to the order with the smallest A_j, to A_1. Areas A and D are then 0. With the *watt* then created, there is an equilibrium between overload and underload while maintaining β. However, this would require removing numerous work operations from the line in our example of Fig. 5.6.

Variant 2

No orders can or should be removed from the assembly line. This may be the case if it is either technically impossible to carry out the activity independently of the main assembly object in pre-assembly or at a supplier. Or the activity is very strongly variance-/option-dependent, an external relocation then increases the complexity costs disproportionately.

In variant 2, the takt time results according to (Moench et al., 2020):

$$T = \max\left\{\max_j a_j/(1+\beta), watt\right\}$$

According to this formula, the takt time $T = watt$ if the spread between the order with the largest workload A_J and the *watt* is less than β. Areas A and D do not exist in this case. Figure 5.7 shows the case where we place the boundary $T^*(1 + \beta)$ at the order with the largest assembly time A_J. In this case, $T > watt$, area D does not exist,

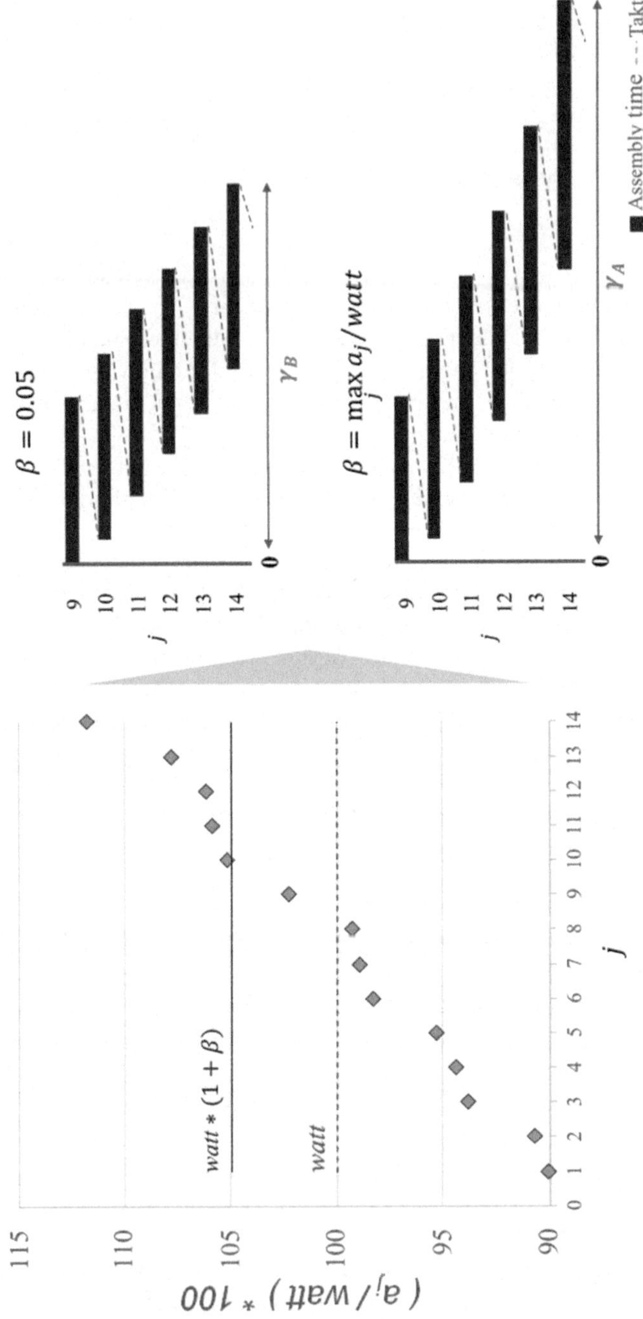

Fig. 5.5 Fendt model 1 workload distribution and worker drift ranges at different β (Moench et al., 2020)

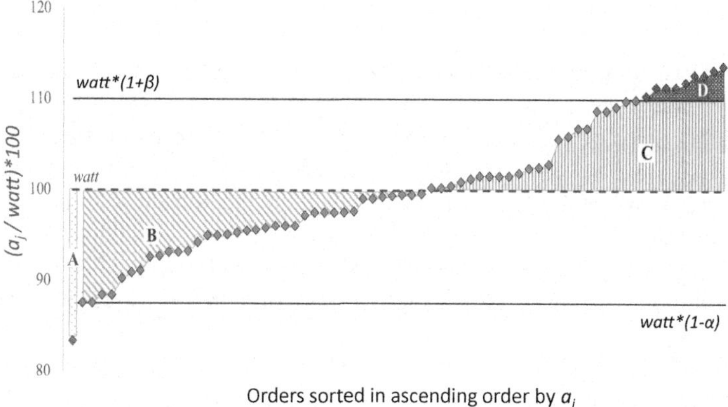

Fig. 5.6 Workload equilibrium for an assembly line with relocatable assembly content and $T = watt$ (adapted from Moench et al., 2020)

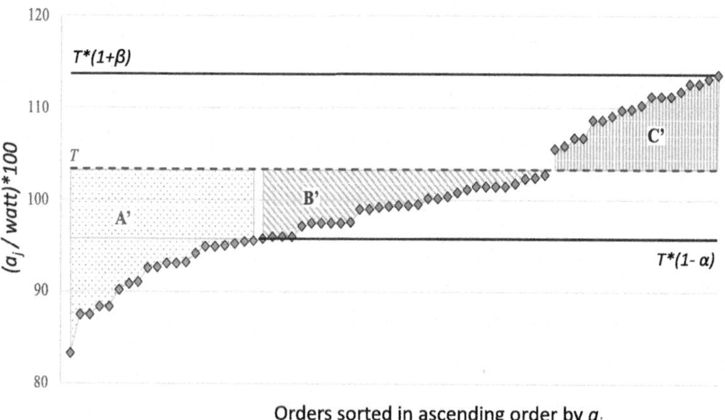

Fig. 5.7 Workload equilibrium for an assembly line with non-relocatable assembly contents and $T = a_J/(1 + \beta)$ (adapted from Moench et al., 2020)

and areas B and C represent a balance between overload and underload. Area A shows the fraction of underload that will not be mastered even in the optimal sequence in our example and will lead to losses. According to our definition in Sect. 2.8.1, these are the takt losses caused by the design of the balancing. Exactly at this point, we can use the variable takt to also master these takt losses of area A or to prevent them from occurring in the first place.

5.3.2 Use of Variable Takt with Work Equilibrium: A Model

In Chap. 4, we showed how two different takt times can be mapped in an assembly line using variable takt. If we apply this to the data in Fig. 5.7, which continues to be the assembly times of a tractor in a part of the Fendt assembly line, then this tractor model could be balanced with two takt times. Since β defines the maximum overtakting, we do not need to fully utilize it in the following example, and a separate overtakt limit can be chosen for each takt time, we denote the overtakt limit per takt time group i as θ_i. Here, $\theta_i < \beta$ always holds. In Fig. 5.8, we eliminate area A and hence the takt losses using two takt time groups; moreover, area B_i is equal to area C_i.

As a perfectly balanced assembly line is assumed, the assembly time a_j of an order j is identical at each station n. In the following calculations, $a_j = A_j/N$ can therefore always be assumed.

This procedure can be described in general terms with the following model (Moench et al., 2020):

a_j, a_r, a_q	Assembly time per station from j, r, q (one worker per station)
i	Index of the takt time group
I	Number of variable takt time groups
j, r, q	Order index
T_i	Takt time of the takt time group i
x_{ji}	Binary variable of the order j to the membership of the takt time group i
β	Maximum overtakt/maximum drift per job

Our model can be used for two objectives. Either (5.1) to minimize the number of takt time groups I *for a* given β (5.1) or (5.2) to minimize the maximum overtakt β for a given number I of takt time groups (5.2).

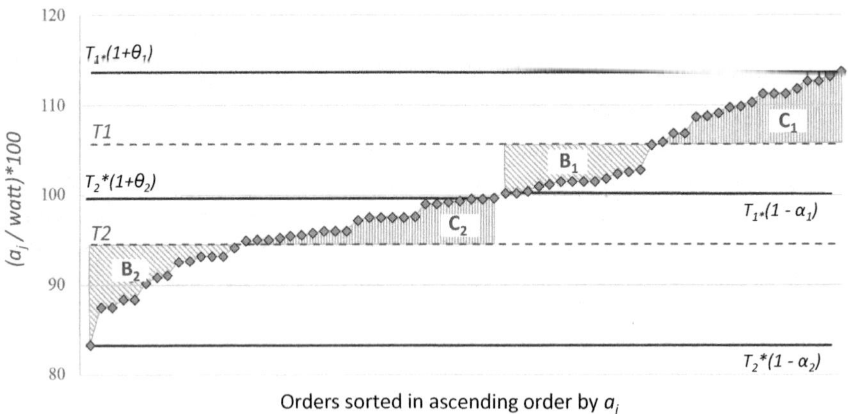

Fig. 5.8 Workload equilibrium for an assembly line with variable takt times (adapted from Moench et al., 2020)

$$\min I \tag{5.1}$$

or

$$\min \beta \tag{5.2}$$

s.t.

$$\sum_{i=1}^{I} x_{ji} = 1 \quad \forall j \tag{5.3}$$

$$a_j * x_{ji} \leq T_i * (1+\beta) \quad \forall i,j \tag{5.4}$$

$$\sum_{q=1}^{Q} A_q * x_{qi} - T_i * \sum_{q=1}^{Q} x_{qi} = T_i * \sum_{r=1}^{R} x_{ri} - \sum_{r=1}^{R} A_r * x_{ri} \quad \forall \mathrm{i} \tag{5.5}$$

$$T_i + \varepsilon \leq T_{i+1} \quad \forall \mathrm{i} \tag{5.6}$$

$$T_i \geq 0, x_{ij} \in [0,1] \tag{5.7}$$

The binary variable x_{ji} ensures that each order is assigned to exactly one takt time group (5.3). If job j is assigned to takt time group i, then $x_{ji} = 1$. Jobs assigned to takt time group i must not exceed the maximum allowable overtakt (5.4). For each takt time group, there must be a balance between overload and underload (5.5), and the areas $\mathrm{B_i}$ and $\mathrm{C_i}$ must have the same value (Fig. 5.8). Overload is caused by all orders $q,\ q = 1\ldots Q$ with their assembly times a_q, which are above the takt time (5.5). Underload is caused by all orders $r,\ r = 1\ldots R$ with their assembly times a_r that are below the takt time (5.5). Formula (5.6) ensures that all takt times are sorted in ascending order. The conditions in line 7 define the non-negativity of the takt times and the binary variables.

Experienced optimizers will immediately recognize the non-linearity in the constraint in line 5. For the solution of this optimization problem, this non-linearity has to be resolved, but for a better readability and understanding of our model, we refrain from doing so here.

Due to the given constraints, the optimization of the models turned out to be a complex mixed-integer optimization problem, which cannot be solved efficiently.

For this reason, a heuristic method based on this model was designed to achieve both goals of (1) minimizing the number of takt time groups and (2) minimizing the maximum overtakt β.

5.3.3 The Variable Takt Time Group Algorithm (VTGA)

The algorithm described in the following can easily be transferred into all common programming languages and was published in the article of Moench et al. (2020). It can be used to analyze data sets with the assembly times of several thousand orders in seconds or to divide them into takt time groups. The algorithm consists of two basic steps. In the first step (Table 5.1), starting from order J with the largest

Table 5.1 Variable Takt time Group Algorithm (VTGA)—Part 1

Algorithm	
Set $\beta, i = 1, j = J$	
Sort a_j in ascending order	
While IDL > 0	
$Q = j$	
$T_i = \max_j a_j / (1 + \beta)$, where $j \notin \Omega_1, \ldots, \Omega_I$	
$watt_i = (\sum_{j'=1}^{Q} a_{j'})/Q$	
If $T_i \leq watt_i$ Then	Work equilibrium
$T_i = watt_i$	
IDL = 0; Exit While	
Else	
While $a_j > T_i$	
$O = O + a_j - T_i$	Overload
$j = j - 1$	
Wend	
While $U \leq O$	
$U = U + T_i - a_j$	Underload
$j = j - 1$	
Wend	
If $O < U$ And $O > 0$ Then	
$j = j + 1$	Equilibrium
$T_i = (\sum_{j'=j}^{Q} a_{j'})/(Q - j + 1)$	
End If	
For $x = 1$ To $j - 1$	
IDL = IDL + $T_i - a_j$	Calculation area A = idle time
Next	
If IDL = 0 Then	
Exit While	
End If	
End if	
$j = j - 1$	
$i = i + 1$	
Wend	
$I^* = i$	

Table 5.2 Variable Takt time Group Algorithm (VTGA)—Part 2

```
34  Set I = I*, j = J
35  While IDL = 0
36      Steps 4-30
37      β = β − ε
38  Wend
```

assembly time A_J, all orders are assigned to takt time groups, taking into account the maximum overload per order β. In step two (Table 5.2), β is minimized for the number of takt time groups found while maintaining the workload equilibrium.

First, β, i, and j are initialized (line 1) and all orders are sorted in ascending order (line 2). Orders are assigned to takt time groups and the number of takt time groups is increased as long as waiting time (area A = idle time) is greater than zero (line 3). In each iteration of the while loop, orders not yet assigned are assigned to a takt time group, where Q stores the index of the largest order not yet assigned (line 4). The takt time T_i of the current takt time group i is set using β (line 5). The current *watt* of all unassigned orders is calculated (line 6) and compared to the current takt time T_i (line 7), if the *watt* is above the current takt time, then the last takt time group is found, the takt time of this takt time group is set to *watt* (line 8) and the entire while loop is terminated (line 9). If the currently calculated *watt* is smaller than the takt time of the current takt time group, then the last takt time group has not yet been found and from line 10 onwards the orders are searched for which still have to be assigned to the current takt time group. In lines 11–14, the orders are searched for whose workload a_j *is* greater than the takt time T_i, their overload is added up, and the area C is calculated. In lines 15–18, the underload of the areas whose workload a_j *is* less than T_i is added up, area B is calculated. Since the orders are sorted in ascending order of workload, area C is calculated first, and then area B is calculated. Since overload and underload are never perfectly balanced, the last order of the current takt time group is taken out again (line 20) if the overload is greater than the underload (line 19). Afterward, the takt time of the current takt time group is recalculated (line 21) or adjusted to the correct value. In the following, the waiting time or idle time (area A) of all orders not yet assigned is calculated (lines 23–25). If the idle time should be the value 0, then the last takt time group was found and the while loop can be terminated (lines 26–28). If this is not the case, the current takt time group index is increased by one (line 31), and j is set to the first unassigned job (line 30). For the following step 2 of the VTGA, the number of takt time groups is stored in I^* (line 33).

In step 1, the takt time groups are calculated taking β into account; here, β is utilized as much as possible. We denote the maximum overtakt of the individual takt time groups as θ_i (see Sect. 5.3.2), by the VTGA the last θ_i, i.e., θ_I, will be significantly smaller than $\theta_1 \ldots \theta_{I^*-1}$ will be. In order to equalize θ_i and to balance the takt time groups, we used step 2 of the VTGA (Table 5.2).

Without going into details, step 2 is explained very quickly. The found number I of necessary takt time groups is kept. In iteration steps β is reduced by a small amount ε and steps 4–30 of part 1 of the VTGA are run through again. This is repeated until the number I of takt time groups is insufficient to reduce the area A to zero (line 35). At the end of step 2, all θ_i are approximately the same size.

5.3.4 Application of the VTGA to a Case Study

An application of the VTGA was shown by Moench et al. (2020), which is based on the data sets described in Sect. 3.2.5. It is a data set of the assembly times of about 450 tractors. Using the example of Fendt tractor assembly, the superiority of the variable takt and the effect of a different number of takt time groups could be shown. It can be seen from Fig. 5.9 that the takt losses (idle time) are greatest at $I = 1$, which corresponds to a fixed takt, regardless of the β. These losses are reduced when variable takt is used and the number of takt time groups i increases, or when β increases.

5.3.5 Limitations of the Method

(a) As described in the limitations of the method from Sect. 5.1, when using data from an assembly line optimized for fixed takt, the full potential of variable takt cannot be extracted. (For a detailed description see "Limitations of the Method" of Sect. 5.1.)

(b) The algorithm assumes a perfectly levelled line balancing, i.e., each station n or each worker w is assigned the identical assembly time a_j at job j. This will never be the case in a real-life application. If the total workload A_j of two orders differs

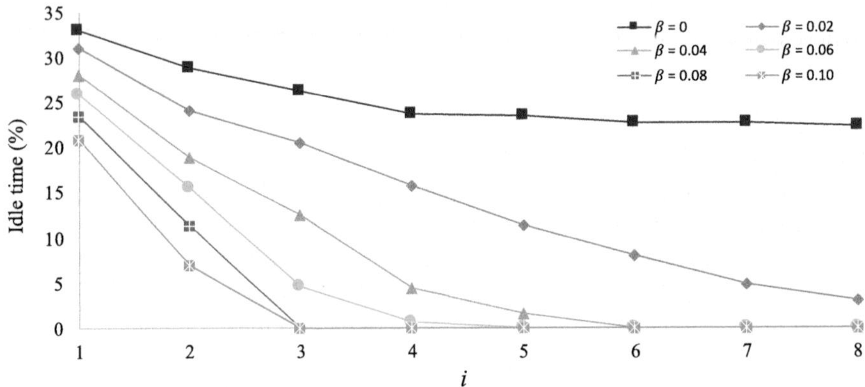

Fig. 5.9 Takt losses (area A, Sect. 5.3.1) as a function of the number of takt time groups and β of the Fendt tractor assembly (adapted from Moench et al. 2020)

due to an option, this time difference will usually be concentrated on certain stations relevant for this option.

(c) Necessary allowance times are not considered but could be represented very simply by $T_{i*} = T_i\,(1 + e_{pa} + e_{fa})$.

(d) By using the VTGA, the segmentation of the takt time groups is carried out in analogy to the descriptions in Sect. 5.2 with Figs. 5.2 and 5.3 exclusively oriented to the assembly time A_j. A previous segmentation according to product groups or the SWS structure existing in the assembly area does not take place. If three product groups $M = 3$ are manufactured in the assembly line and the application of the VTGA results in three takt time groups $I = 3$, then 3*3 = 9 line balances (9 different SWSs per station) could become necessary. This means a high administrative effort and a significantly higher complexity for the workers on the line.

(e) If the lower limit of a takt time group in the VTGA falls on a plateau of orders (several orders with the same A_j), then the VTGA separates these orders and can assign them to different takt time groups. In the logic of the VTGA this is correct, as it ensures the equilibrium between overload and underload. However, for a practical application, this case is not feasible because orders with the same A_j are assigned to different takt time groups. This could easily be resolved by assigning all these A_j *to* the next takt time group T_{i+1} in an additional step and then redetermining the value of the takt time group T_i.

(f) The method described here is based on the use of past data, thus its conclusions for future production are tied to a similar model-mix and option distribution.

5.3.6 Conclusion VTGA

The VTGA represents a first quantitative method to determine takt time groups, but a direct practical implementation proved to be very difficult due to the limitations shown in Sect. 5.3.5. The idea of workload equilibrium and the restriction of the maximum overtakt by β however, have a very high practical relevance. If the approaches from Chap. 2 and the procedure from Fig. 5.4 are combined with the findings from the VTGA, an improved procedure for the segmentation of takt time groups in existing assemblies can be developed, which we present in the following chapter.

5.4 VTGTP: Variable Takt Time Group Transformation Process

With the VTGTP (*V*ariable *T*akt time *G*roup *T*ransformation *P*rocess), we extend and modify our qualitative procedure from Sect. 5.2. The VTGTP (Fig. 5.10) is intended to be a generally applicable process with which an assembly (line) is analyzed for the conversion to variable takt and the most important parameters are determined. In the first step, we divide the production into relevant study areas, the assembly areas to be transformed must be able to convert to a continuous flow motion with adjustable

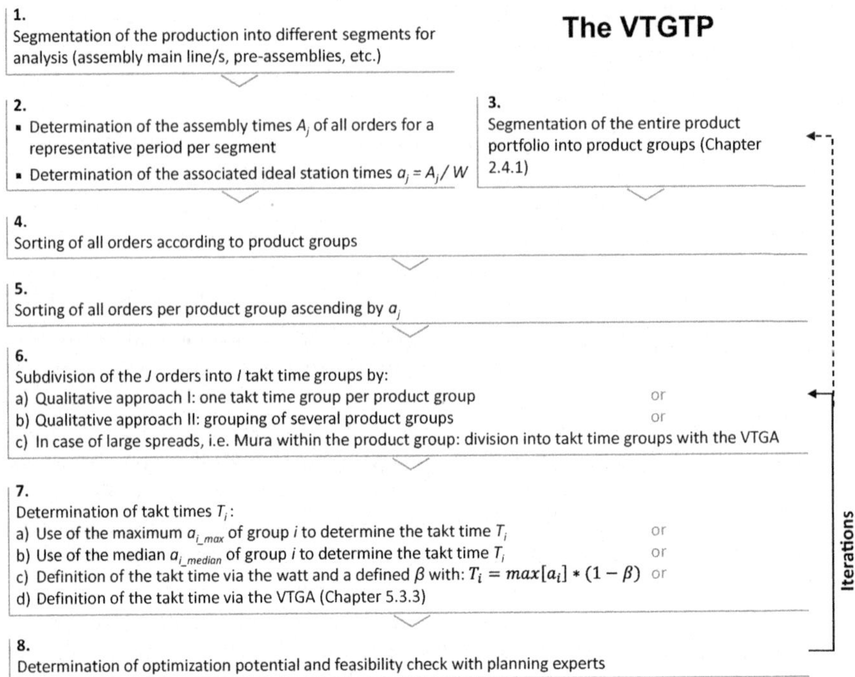

Fig. 5.10 The VTGTP—Variable Takt Time Group Transformation Process (adapted from Hiller, 2018)

launching intervals. Subsequently, all total assembly workloads A_j per section of the assembly line under consideration are calculated for a representative period of time. Representative in the sense that the distribution of the different products and option variants can be used for the future planning horizon under consideration. Using the table from Sect. 2.4.1, the entire product portfolio is divided into product groups. In step 4, all orders are first sorted according to the product groups and then, in step 5, sorted in ascending order according to their a_j. This is followed by the segmentation of the orders into I takt time groups. This can be done qualitatively by visualizing the trend lines of the a_j or by the VTGA. With the determination of the takt times T_i, the degree of tension (stress) in the assembly is essentially determined. If the maximum a_j of a takt time group is used for this purpose (maximum takt), this results in the lowest degree of tension in the line balancing and in the execution of the assembly activities, but also the lowest productivity. Nevertheless, when introducing variable takt, it can make sense to chose this step first in order to win over all those involved in the implementation of variable takt. In the later course and after successful introduction, the takt times can then be further adjusted. Based on our practical experience, steps 3 to 7 are run through several times, as the results should be discussed and checked for feasibility with the management and planning experts of the assembly concerned.

5.5 Case Study: Canyon Bicycles

In the following section, we would like to describe the application and introduction of variable takt using a practical example. We were able to win the company Canyon Bicycles from Koblenz, Germany, for this. After a short description of the company and the bicycle assembly, we apply our VTGTP to determine and classify the takt time groups. All used data represent only a part of the assembly of Canyon Bicycles, the main assembly line, to the other production sections no statements can be made, in order not to publish competition-relevant data.

5.5.1 Canyon Bicycles GmbH

Started out of a Koblenz garage as "Radsport Arnold," the company, which has been operating under the name Canyon Bicycles GmbH since 2002, is now one of the world's leading manufacturers of road, mountain, triathlon, fitness, and urban bikes as well as kids' and e-bikes. Canyon Bicycles stands for passion for cycling as well as numerous award-winning innovations and works with the world's numerous athletes. Around the globe, Canyon Bicycles is appreciated by cyclists for innovative products, leading technologies, highest quality, reliable service and, last but not least, the multiple award-winning design. In 2008, the company generated more revenue in its international sales markets than in its core German market for the first time. As a "Direct-2-Consumer" brand, Canyon bikes are available worldwide exclusively online at canyon.com.

Canyon Bicycles currently employs around 1000 people in a variety of business areas including development and design, marketing and sales, production and logistics, consulting, and service. Canyon Bicycles's workforce grew by around 20% in 2020. The company even increased its global sales across all companies by around 37% to just over 415 million euros in the 2020 financial year.

The assembly of the bikes is carried out at the Koblenz site in a hall with 9200 m^2 in one shift (Fig. 5.11). Across all bicycle types, several tens of thousands of units are assembled each year, with a strong upward trend. The Corona years 2020 and 2021 in particular have brought about a significant surge in demand. Assembly is organized in the structure of an intermittent flow production (see Sect. 2.2) with a takt time $T_{Fix} = 80$ s. Since the process time of the largest product is $a_j = 60$ s, a balancing efficiency of $E_{AL} = 0.85$ is aimed for and an allowance time of $e_a = 0.05$ is to be covered, the fixed takt time $T_{Fix} = 80$ s is appropriately selected (see calculation of takt time in Sect. 4.4.3). The production units are not moved continuously, but after the takt time has elapsed, all units are conveyed synchronously to the next station via a light signal. The conveyor system is a friction-based overhead conveyor with independent drive units. Although one of the prerequisites for using the variable takt, continuous flow operation, has not yet been implemented, this can be achieved with the existing conveyor system. The drive units can be controlled independently of each other, it is therefore possible to vary the spacing of the products, and the system can be operated in continuous flow mode up to a speed of $V_{CS} = 0.03$ m/s.

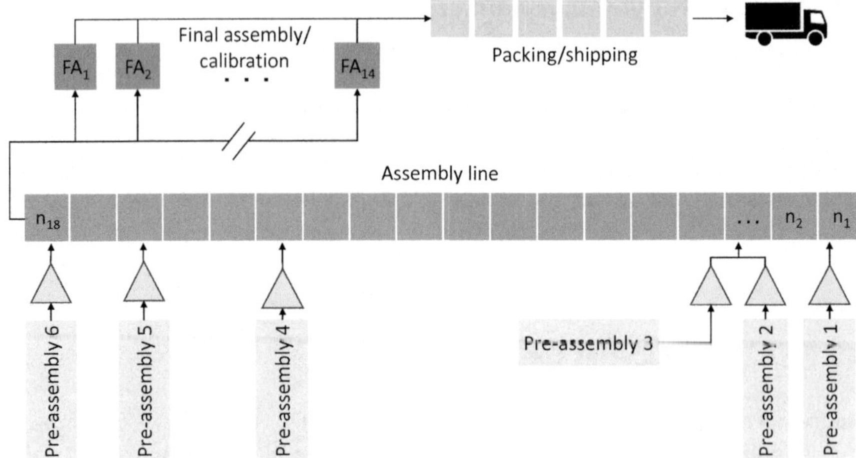

Fig. 5.11 Assembly structure Canyon Bicycles Koblenz site

Fig. 5.12 Assembly line—Canyon Bicycles Koblenz site

When designing conveyor systems for assembly, not only the maximum speed is an important parameter, but also the minimum speed at which the system can still represent a steady flow operation are important parameters that are very often forgotten as a requirement in the specifications when designing the system (5.11).

The main assembly line (Fig. 5.12) consists of a total of 18 stations $N = 18$, six different pre-assembly stations are decoupled from the line by means of buffers and connected upstream. The takt time of the main assembly line is $T_{Fix} = 80$ s. The assembly line is followed by an M1 matrix structure (cf. Sect. 7.1), the final assembly and adjustment, in which the gears and brakes of the bicycles are adjusted and final quality checks are carried out. Finally, the packaging of the products takes place in a line with six stations.

5.5.2 Segmentation of the Production Program into Takt Time Groups Using VTGTP

Step 1: Segmentation of the production into different segments for analysis
From Fig. 5.10, we select the main assembly line with its 18 stations as the area of investigation. We initially exclude pre-assembly, final assembly, and adjustment section as well as the packaging area from our considerations. These areas are not organized as flow assembly.

Step 2: Determine all assembly times A_j
At the time of the case study, Canyon's product and production portfolio includes more than 100 differently configured bicycles, the product range is being continuously expanded, and a significant expansion of customization is targeted in the coming years. We consider the production period of several months. The bike with the highest demand was produced more than 5000 times during this period, the one with the lowest demand came to a quantity of 30. The A_j were determined for all orders.

To better derive the ideal takt in the later visualization, we first divide the A_j of all products by the number of workers and obtain a_j. Since one station in the assembly line is not occupied and all other stations are occupied by one worker, $N \neq W = 17$. We calculate in this case: $a_j = \frac{A_j}{W.}$

Step 3: Segmentation of the entire product portfolio into product groups
For our investigation, we first divided the entire product portfolio into four product groups; for this, we use our table from Sect. 2.4.1. The four product groups (Fig. 5.13):

- Differ significantly in their design structure
- Are perceived within the company as independent product groups
- Each has its own BOM structure
- The products in the product group differ in option variants

Product group Road – $\mathbf{P_{Road}}$ Road bikes, such as the Canyon Ultimate, are designed specifically for road and street cycling. They are characterized by a lightweight construction (often carbon, narrow tires, etc.) and are reduced to the parts necessary for cycling (i.e., no fender, rack, etc.).

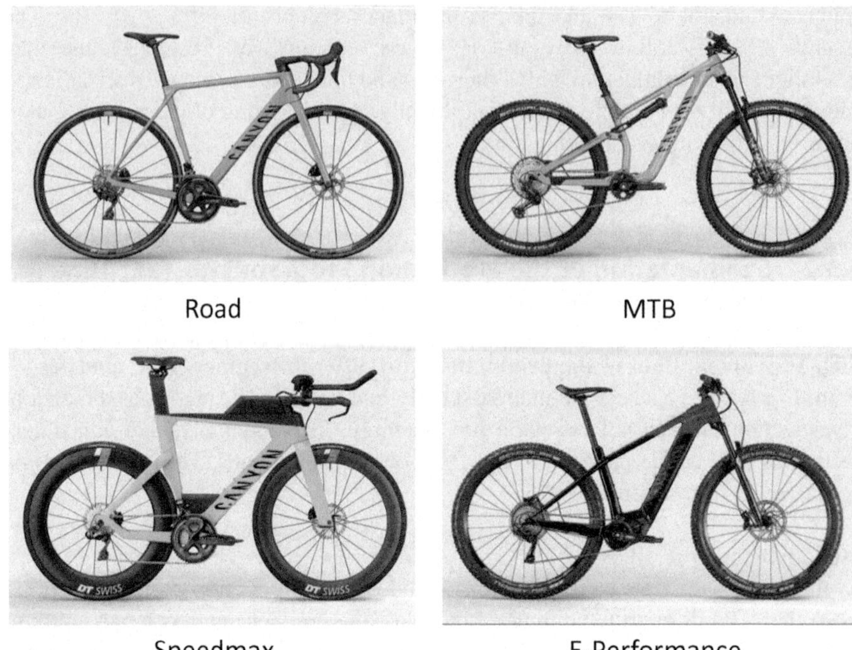

Fig. 5.13 The four product groups Road, MTB, Speedmax, and E-Performance

Product group Mountain Bike – P_{MTB} Mountain bikes, such as the Canyon Neuron, are bikes that are designed for use off paved roads. They are characterized by a more stable frame and special suspension elements on the fork and/or rear triangle.

Product group Speedmax – P_{SPM} The Canyon Speedmax is a bike designed for triathlon sports and competitions on the road. The geometry, integrated components, and the specially developed aero handlebars allow for a particularly aero-dynamic riding position. Thus, P_{SPM} differs significantly in the basic structure to P_{Road}.

Product group E-Performance Bikes – $P_{E\text{-}Performance}$ The category "E-Performance Bikes" includes bikes with motorized support that are suitable for sporty use off-road and on the road—that is, e-mountain bikes or also e-race bikes.

Step 4 and 5: Sorting of all orders according to (1) product groups and (2) ascending by assembly workload a_j

The trend line is shown in Fig. 5.14.

Step 6 and 7: Qualitative segmentation of the orders into takt time groups and determination of the takt times T_i

Figure 5.14 shows that the different bikes/orders of the E-Performance product group do not differ in their a_j of the main assembly line. The option differences

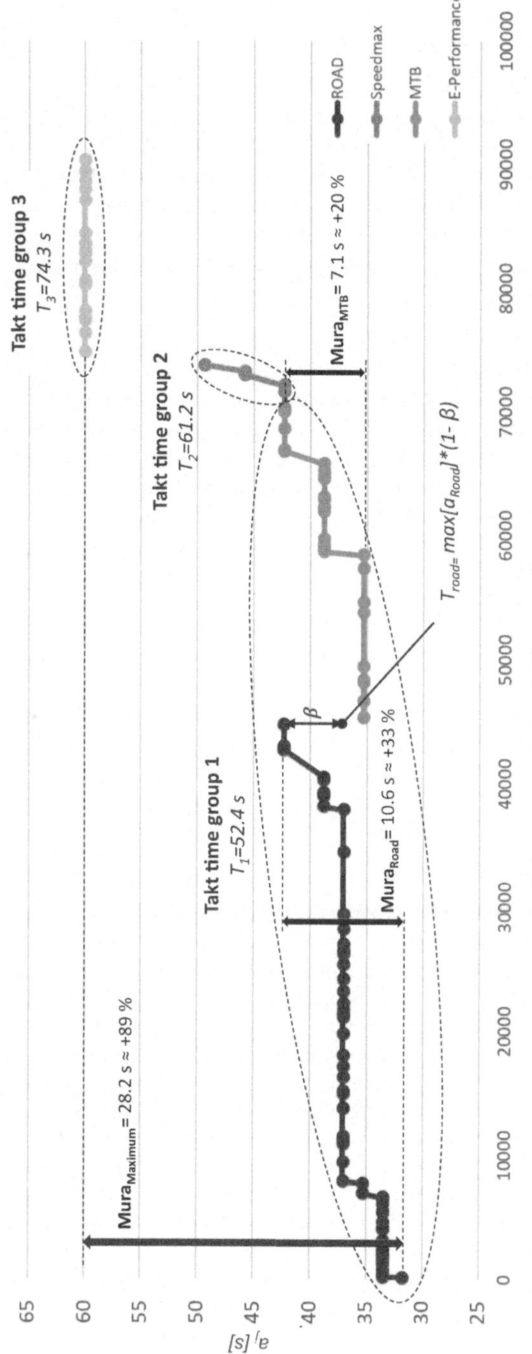

Fig. 5.14 Orders of Canyon Bicycles sorted according to (1) product group and (2) assembly line workload a_j

have already been shifted to the pre-assembly stages or are in the area of different components that take the same assembly time. The orders in the Speedmax, Road, and MTB product groups show differences in assembly time due to different option variants. Mura in the a_j in the Road group is 10.6 s, which corresponds to approximately 33%. Mura in the a_j in the MTB group is 10.6 s, this corresponds to approximately 20.2% (Fig. 5.14). Mura between the smallest and largest order is 28.2 s and thus +89% (Fig. 5.14), a large value for an assembly line in the fixed takt, this is exactly where the variable takt will show its advantage.

In Sect. 3.2.5, we used the same system to show the A_j and a_j curves of the Fendt tractors in the main assembly line. Compared to these representations, Fig. 5.14 shows clearly visible discontinuities, although significantly more orders are represented. The number of options and their possible combinations are many times greater for a tractor than for Canyon Bicycles, resulting in "only" a little over 100 different bikes (colors excluded). In addition, the total assembly time of the tractor is a factor of >100 larger and much finer graded. This leads to a different "look" in the gradients of the a_j between bike and tractor.

It is immediately obvious to divide the products of the E-Performance product group into a separate takt time group. Since the respective a_j did not differ in the product group, the takt times can also be determined directly using the formula from Sect. 4.4.3. $E_{AL} = 0.85$ is selected as the desired line efficiency, and the allowance times are set to $e_{pa} + e_{fa} = 0.05$.

$$\text{E} - \text{Perfomance}: T_3 = \frac{A_{E-Performance}}{W \cdot E_{AL} \cdot (1 - (e_{pa} + e_{fa}))} = \frac{a_{E-Performance}}{E_{AL} \cdot (1 - (e_{pa} + e_{fa}))}$$
$$= \frac{60}{0.85 \cdot (1 - 0.05))} = 74.3\ s$$

The segmentation into takt time groups and the definition of the takt times for the other product groups are not so trivial to define. Together with the production management of Canyon Bicycles, it was decided to assign the product groups Road and MTB to a common takt time group in the first phase of the implementation and to define the takt time via the maximum variant (for reasons, see the following step 8).

$$\text{Road and MTB}: T_1 = \frac{\max\left[a_{Road}; a_{MTB}\right]}{E_{AL} \cdot (1 - (e_{pa} + e_{fa}))} = \frac{42.35}{0.85 \cdot (1 - 0.05))} = 52.4\ s$$

A separate takt time group was selected for the Speedmax product group, which is the smallest by volume, because the variants of this product group differ significantly from the Road and MTB product groups. The maximum a_j is also used to determine the takt time (for reasons see step 8).

$$\text{Speedmax} : T_2 = \frac{\max\left[a_{Speedmax}\right]}{E_{AL} \cdot (1 - (e_{pa} + e_{fa}))} = \frac{49.4}{0.85 \cdot (1 - 0.05))} = 61.2\ s$$

Notes

- *If the assembly line is successfully converted to continuous flow operation and variable takt, the balancing efficiency could be continuously increased. Further productivity increases are thus achieved.*
- *There are six different a_j in the Road product group* (Fig. 5.14). *After a successful introduction of continuous flow operation and the new takt system, the WATT method* (Sect. 2.6) *could be used to master the option variants for this product group in addition to variable takt. We first propose a $\beta = 0.15$. This would allow the takt time to be selected for the Road product group* (see T_{Road} in Fig. 5.13):

$$T_{Road} = \frac{\max\left[a_{Road}\right] * (1 - \beta)}{E_{AL} \cdot (1 - (e_{pa} + e_{fa}))} = \frac{42.35 * (1 - 0.15)}{0.85 \cdot (1 - 0.05))} = 44.6\ s$$

- *It is easy to see that the products in the MTB product group have only three different a_j. The products of the MTB group could thus be divided again into three separate takt time groups in order to realize further productivity increases. In this case, the entire variance in the assembly line would be mastered by the variable takt, and the use of the WATT would not be necessary* (Fig. 5.14).

Step 8: Determination of optimization potential and feasibility check with planning experts

The last step of our VTGTP is a validation with the responsible planning experts and the assembly management. The results of steps 3–7 documented here are the result of several iteration loops. At the outset of our analysis or consideration of converting Canyon Bicycles' assembly to variable takt, we wanted to realize the full productivity potential. By setting $E_{AL} = 0.85$, putting the Road and MTB product groups together into a T_1 takt time group and defining the takt time via the maximum variant, the full productivity potential is certainly not achieved in the first step. Converting the entire line to continuous flow operation and using variable takt times initially represents a very big change for the workers concerned—a noticeable increase in the level of tension. In order to take the affected employees along with them in the changeover process, not to overburden them and to meet the high-quality demands on assembly, the parameters described were chosen for the first implementation step.

5.5.3 Potentials for Increasing Efficiency when Switching to Variable Takt Times

When an assembly line is converted to variable takt, a wide range of advantages arise; we have described ten of these advantages in detail in Sect. 4.5. Potential productivity increases can be derived as follows.

The fixed takt time of the assembly line is 80 seconds. In the current intermittent line operation, 7 s are required for the simultaneous transport of the assembly object carriers to the next station. If converted to continuous flow operation, these 7 s can be filled with assembly content or the line speed could be increased by this amount. The potential for an 8.75% increase in productivity results from this. The total potential productivity increase with our set parameters from Sect. 5.5.2 can be seen in Table 5.3.

With demand for bicycles surging in 2020 and 2021 due to the Corona pandemic and the market success of Canyon Bicycles, 29% more bicycles could be assembled with the same assembly line run time.

In addition to these productivity and output increases, Canyon Bicycles could follow any change in customer demand. If only products of the E-Performance group are in demand or only of the Road group, they could be assembled in build-to-order. It should be noted that the volume produced per time unit adjusts to the product-mix with variable takt (Sect. 4.4.2).

After the successful conversion to a continuous flow assembly and the introduction of variable takt times, the following measures could further increase the productivity of the line:

- E_{AL} should be successively increased to 0.95, through an improvement in the line balancing.
- e_a is left at 0.05.
- As described in Steps 6 and 7 of the VTGTP, the WATT method with a $\beta = 0.15$ could be used for the Road product group.
- The MTB product group is divided into three takt time groups.

With these measures, productivity could be increased by 43% compared to the fixed takt.

Table 5.3 Potential productivity gains with the introduction of variable takt on the Canyon Bicycle main assembly line

	Share of total volume (%)	T_{FIX} [s]	T_i [s]		Increase productivity (%)
Road	50	80	T_1	52.4	35
MTB	28	80	T_1	52.4	35
Speedmax	4	80	T_2	61.2	24
E-performance	18	80	T_3	74.3	7
					29

5.5.4 Design of the Assembly Line: Distances and Line Speed

We have created the most important parameters for the introduction of variable takt, but these must also be "physically" implementable in the current assembly line. The following framework conditions were set:

- An extension of the existing assembly line is not permitted.
- Changes in the carrier/conveyor system should be kept to a minimum.
- The station lengths are to be maintained.
- The conveyor speed should be selected as low as possible to make it easier for employees to switch from stationary assembly to continuous flow assembly.

In Sect. 4.4.4 we presented a procedure to determine the distances of the takt time groups and the station and line length. In order to keep the total line length as short as possible and the line speed V_{CS} as low as possible, the orders of the takt time group with the lowest takt time T_1 should be given the shortest possible spacing. The frames of the bicycles are transported parallel to the flow direction on the goods carrier. The maximum length of the frames is 1.3 meter (Fig. 5.15), the minimum distance between two units is 0.5 meter, resulting in a distance of $D_1 = 1.8$ m for takt time group T_1 and a distance of $D_3 = 2.55$ m for T_3. Since the station length is 2.65 m, no parallel work is necessary even for the products of the E-Performance group (see Table 5.4), i.e., closed station boundaries can be realized. With the probabilities of occurrence as weights and launching intervals from Table 5.4, the result is an average distance of $D_{VarTakt\_mean} = 1.95$ m, which is 26.4% less than the distance in the fixed takt $D_{Fix} = 2.65$ m. Thus, approx. 26.4% or 5 additional goods carriers are required in the main assembly line. It can be seen from Fig. 5.11 that the front and rear wheels are assembled in the last two stations. In these stations, the bicycle is rotated 90 degrees in the goods carrier; without this capability, a $D_1 = D_{min} = 1.8$ m would not be possible.

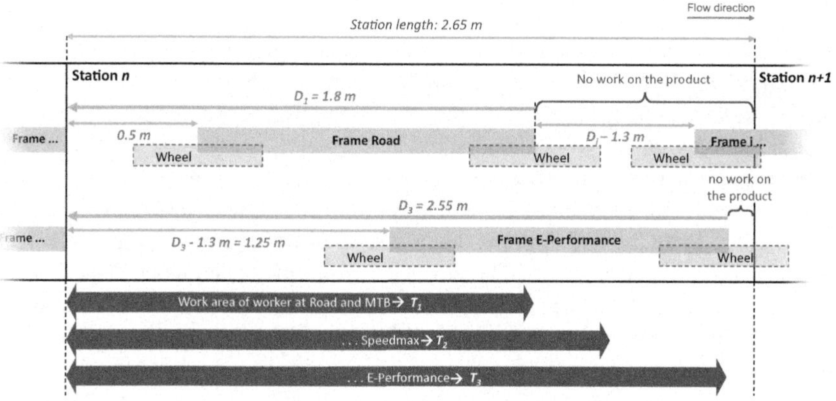

Fig. 5.15 Layout station, cubature frame, distances variable takts of the assembly line Canyon Bicycles

Table 5.4 Distances, line speed and proportion of parallel work with variable takt Canyon Bicycle

	Share of total volume (%)	T_i [s]		D_i [m]	Speed assembly line V_{CS} [cm/s]	Proportion of overlapping work (parallel work) (%)
Road	50	T_1	52.4	1.8	3.4	−32.1
MTB	28	T_1	52.4	1.8		−32.1
Speedmax	4	T_2	61.2	2.10		−20.7
E-performance	18	T_3	74.3	2.55		−3.7

We calculate the speed of the assembly line with the formula from Sect. 4.4.3.

$$V_{CS} = \frac{D_i}{T_i} = 3.4\frac{cm}{s}$$

A line speed of 3.4 cm/s is lower than that of a classic automobile assembly, but in this case, there is no relative speed between the worker and the assembly object due to the use of an apron conveyor. At Canyon Bicycles, the bicycle is transported in an overhead conveyor, and the worker must walk D_i continuously. In total, it is a little more than 1 km per working day, an average value for an assembly worker, from measurements in the Fendt tractor assembly we know that their workers cover significantly longer distances on average.

5.5.5 Conclusions and Further Steps for the Implementation of Variable Takt at Canyon Bicycles

Further technical conditions must be clarified for implementation. Is the conveyor capable of displaying a constant speed of 3.4 cm/s? The conveyor control system must be adapted for variable distances. The layout of the individual stations must be redesigned for continuous flow operation. The MRP and ERP system must be reconfigured for a variable daily output that depends on the model-mix. One of the most important tasks for a successful implementation will be a well communicated and accompanied change process for the employees. The new assembly system must be coordinated with the works council.

With the help of this short case study at Canyon Bicycles, VTGTP was used to show how variable takt could be integrated into an existing assembly line. The potential productivity and volume gains are enormous. Due to the high Mura between different bikes, Canyon Bicycles' assembly is an ideal candidate for using variable takt. After the successful introduction of variable takt and continuous flow assembly, the productivity gains could be significantly expanded again, as we describe at the end of Sect. 5.5.3.

References

Hiller, A. (2018). *Erforschung der Eignung variabler Auflegeintervalle für Variantenfließmontagen der Automobilindustrie*. Munich University of Applied Sciences.

Moench, T., Huchzermeier, A., & Bebersdorf, P. (2020). Variable takt time groups and workload equilibrium. *International Journal of Production Research*. https://doi.org/10.1080/00207543.2020.1864836

6 Design-for-Takt and the Ideal Flow Assembly

Already in the introduction, in Sect. 1.4, we showed that the greatest preventive effect for reducing manufacturing costs can be achieved in the product development process. A finding that can certainly be shared by all experienced managers and planners of an assembly. As model-mix assemblies of type 3 expand, the solution space of Mixed-Model Assembly Design (MMAD) for effective and efficient assemblies becomes smaller, and the entire boundary curve (Fig. 6.1) is shifted to the upper right. The task of Product Design for Mixed-Model Assembly (PD for MMA) is to increase the solution space again. In short: to shift the boundary curve to the lower left in Fig. 6.1. The well-known concepts of Design-for-Assembly (DFA) and Variant Management specifically aim at reducing manufacturing costs, mainly by reducing manufacturing and assembly efforts, quality costs, and variance. We describe these approaches only briefly in the following chapter because they are not the focus of our considerations. Our focus remains on mastering variance in assembly. For this purpose, Swist (2014) presents for the first time the concept of what we call Design-for-Takt, which we take up and extend with our own considerations, experiences, and analyses. The purpose of Design-for-Takt is the reduction of utilization losses (Sect. 2.8), for this reason, we extend the labeling of the Y-axis of Fig. 6.1 by this focus.

6.1 Design-for-Assembly

DFA (Design-for-Assembly) describes a process in which individual parts, assemblies, or the entire architecture of the product are designed in such a way that it leads to a

- simpler (reduction of qualification effort, operating resources),
- more efficient (reduction of assembly workload),
- more effective (reduction of error possibilities)

P. Bebersdorf, A. Huchzermeier, *Variable Takt Principle*, Management for Professionals, https://doi.org/10.1007/978-3-030-87170-3_6

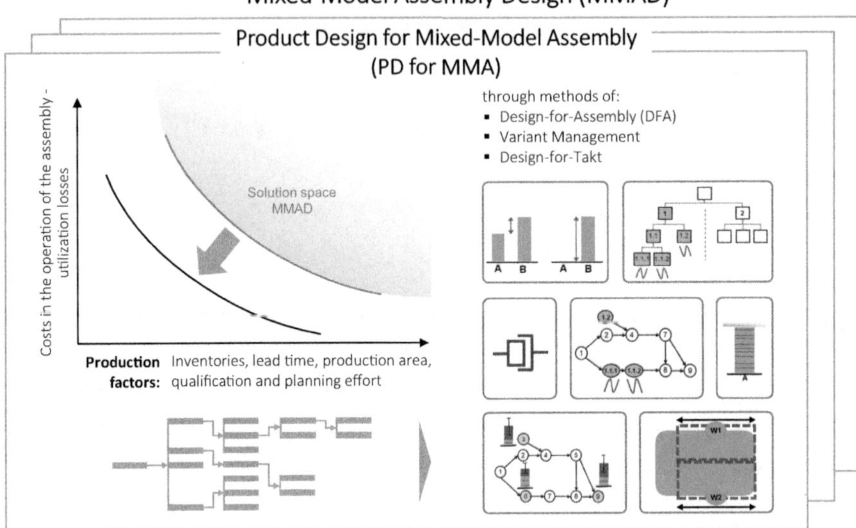

Fig. 6.1 Product design for mixed-model assembly (PD for MMA), adapted from Swist (2014) and Kesselring (2021)

assembly (Andreasen et al., 1988). Here, on the one hand, interventions on individual components usually have a short-term and limited effect (Swist, 2014). Modifications on the entire product architecture, on the other hand, have the greatest lasting impact in reducing manufacturing costs, but can usually only be implemented for downstream products (Swist, 2014). Numerous approaches to implementing DFA are available and described in the existing literature. Even in a VDI (Association of German Engineers) guideline, the VDI 2221, a design process is described, which should lead to a product design that is suitable for assembly. Boothroyd (2005) is one of the best-known and most successful researchers in this field.
According to Boothroyd (2005), the two factors determining assembly times are:

- The number of necessary parts and
- The number and effort of (the three) basic activities, i.e., handling, joining, and fastening

As part of his systematizations, Boothroyd developed the DFA Index, which has been adopted by a large number of other publications:

$$E_{ma} = \frac{N_{min} \cdot t_a}{t_{ma}}$$

with

E_{ma} DFA index
N_{min} Theoretical lowest number of parts

t_a Basic assembly time of a part
t_{ma} Assembly time for a part

For details on how to determine the theoretically lowest number of parts and for further details on the DFA Index, please refer to the publications of Boothroyd, whose approach has been implemented for many years by the company "Boothroyd Dewhurst Inc." (www.dfma.com). From the DFA index, it can be seen that the number of parts to be assembled and the ease of execution of the three basic assembly activities are the decisive factors of the assembly effort—with the number of parts predominating. In addition, Boothroyd (2005) proved that the quality of assembly can be increased as the DFA index increases.

In the DFA approach, no elements can be identified that reduce the losses in the assembly process of a mixed-model assembly; utilization losses (Sect. 2.8) are not addressed.

6.2 Variant Management

In Variant Management, the focus is on reducing the complexity of products (Wiendahl et al., 2004). The aim is to achieve maximum external variance (from the customer's point of view) with the minimum possible internal variance (from the manufacturer's point of view). By reducing the number of variants in the product portfolio, operating resources can also be simplified and investments avoided.

The most comprehensive overview of approaches to Variant Management is given by Wiendahl et al. (2004) in their work, which we summarize in Table 6.1.

In none of the known approaches to Variant Management do utilization losses in the form of takt or model-mix losses play a role.

6.3 Design-for-Takt

In their work, design engineers are confronted with a multitude of wishes or demands from different areas of a company. Tension between different interests arises. In a customer-centric company, on the one hand, marketing and sales usually make the initial demands, so existing features should be changed or new options, for which sales and marketing forecast a positive demand, should be created. On the other hand, colleagues from the purchasing department will certainly demand the use of existing parts, want as little variety of parts as possible, or basically demand the use of materials and designs that entail the lowest possible procurement price.

The initial task of the designer is to create the desired function in the space provided. Furthermore, the designer has to keep to his/her project budget, which is why he/she usually only strives for small changes or additive elements to existing design solutions. In practice, this is often referred to as "onion design"—in order to save development costs and minimize the risks of a new design, changes to existing systems or new content are incrementally designed—"onion skin by onion skin". The software or physical design literally gets wider and wider with each new

Table 6.1 Approaches to Variant Management (adapted from Wiendahl et al., 2004)

1. Standardization through the use of norm and identical parts	The use of standard parts limits the variety of solutions available to designers. In this way, the degree of reuse and the interchangeability of components can be increased. The number of components with variance elements can be reduced by concentrating the variance content of existing parts. The remaining basic functionality is mapped with identical parts. This principle allows the point of origin of the variant in the assembly process to be moved as far to the end as possible (from the perspective of the precedence graph).
2. Integral and differential construction	If several functions can be integrally combined in one component, a larger number of identical parts is created. Also in the differential design, the repetition frequency of components is to be increased, this is done by decomposing the variance scope into several standardized parts.
3. Modularization	Modularization involves breaking down basic and variance functions of products into modules that are as independent of each other as possible. These modules should be freely combinable within the respective product or among different products. In this way, the internal complexity should be minimized and the external variance maximized. Successful modularization can also move the point of origin of the variant near the end of the assembly process.
4. Modular toolkit system and platforms	Both modular toolkit systems and platforms are based on the concept of modularization; modular toolkit system, in particular, can be seen as a general implementation of modularization. Variant-specific modules are mounted on one or more variant-free basic modules (Swist, 2014). Two of the best-known historical platform concepts are the Modular Transverse Toolkit (in German MQB) and the Modular Longitudinal Toolkit (in German MLB) of Volkswagen, or their electronic successors PPE (Premium Platform Electric) and MEB (Modular E-Drive Toolkit). The Scalable System Platform (SSP) is the future platform for electronics, software, and drive technology on which Volkswagen will base all future models in the Group. In the context of increasing electrification in automotive engineering, joint platform concepts are also growing strongly between different manufacturers. "The component with the largest volume of temporally stable units is the platform" (Wiendahl et al., 2004, p. 48), which can usually be used over several product life cycles.
5. Packages	Without changing the product structure, package formation makes it possible to reduce the number of final product variants. This is because the number of option variants that can be combined is restricted in package formation (Rapp, 1999). The contents of one package typically lock the contents of the other package.

(continued)

Table 6.1 (continued)

6. Variants in the precedence graph, late variant formation	The aim of this approach is to plan and design the variants so that they can be assembled at a very late stage in the assembly process, in the precedence graph respectively. The variant creation point is placed far towards the end of the assembly process. The trend towards standardized, neutral hardware and customer-specific software are current reactions. If there are many degrees of freedom in the precedence graph, variants can be selectively bundled and moved to pre-assembly or hybrid stations (Sect. 8.2).

innovation round—or "onion skin." In terms of manufacturing costs, this usually results in suboptimal designs.

In the context of DFA and Variant Management, the demands of assembly on the development process have mostly been related to the reduction of assembly times and the reduction of variants to be mastered. In Sects. 1.2 and 4.1, we described recent developments that force assemblers to manage more variance in and on products. With an increase of model-mix assemblies, especially of type 3, further requirements of the assembly have to be added to the design of the products. The Design-for-Takt is therefore not in competition with DFA and Variant Management, but should be a useful addition from the point of view of a clocked flow assembly.

6.3.1 Objectives of Design-for-Takt

Already at the beginning of this book, in Sect. 1.2.4, we showed the main challenges that arise when model-mix assemblies of type 3 wants to master variance:

1. Minimize utilization losses in assembly operations
2. Reduce change and maintenance costs for adjustments to volume, product-mix or design
3. Reduce investment when initializing or modifying assembly

While the original Design-for-Takt concept under Swist (2014) focused only on reducing utilization losses, we extend this to include the goals of reducing change costs and capital expenditures.

Minimization of utilization losses

Design-for-Takt represents a new approach to solving four existing problems in minimizing utilization losses:

- The *existing design* of products already represents a *limited solution space*, and the reactive processes that then follow to reduce takt and model-mix losses thus become or are very costly processes and reduce utilization losses at the expense of other target values such as floor space requirements, inventories, lead times, qualifications, etc. (Swist, 2014).

- In the design of the products, *effects on future utilization losses are unknown, so other targets* such as construction space, functional scope, parts, and development costs *dominate.* A levelled assembly workload of the products to be assembled and their options are not in focus.
- The *flexibilization of assembly precedence graphs*, through a flexibilization of constructive precedence relationships, is *not a recognized goal.* The concentration of variance or the targeted distribution of variance in assembly is thus not possible or only possible with great difficulty.
- Coherently designed scopes are usually cost-optimal in terms of their part costs and total assembly time due to minimal interfaces and thus a minimized number of parts. *However,* larger *assembly content that cannot be divided further* in relation to the takt time *severely restricts the degrees of freedom in the line balancing* and thus increases takt losses.

Reduction of modification costs
Due to changing framework conditions for assembly, such as changes in volume, product-mix, product changes, or efficiency increases, assemblies have to adapt continuously. Line balancing has to be adjusted at regular intervals, resulting in the following two problems, which lead to increased change costs:

- *Rigid precedence relationships severely restrict the degrees of freedom in rescheduling*; making these relationships more flexible reduces the complexity in the rescheduling process.
- *Non-divisible, large assembly scopes* (in relation to the takt time) severely *restrict the degrees of freedom* for re-balancing.

Reduction investments
The design of the products significantly sets the framework conditions for the required production area and equipment of the assembly. The cubature and weights or weight distribution of the entire product, but also of the individual parts and modules, must be handled in the assembly:

- If *many different components* are required to map the variance, *production areas must be reserved for this purpose* or complex JIS processes must be installed.
- If these components cannot be transported in standardized load carriers, *special load carriers or even special transports* are necessary.
- A determining factor for the required assembly area is not only the volume to be produced, but also the total assembly workload of the individual products. *If the workload is concentrated* on a specific point along the cubature of the product and the degrees of freedom in the precedence relationships are very low, the assembly *activities can only be carried out serially*—the assembly literally "drags in length."

- A *large number of different assembly elements* (e.g., screws) or assembly parameters (e.g., torques for screw connections) require a larger number of or increased *complexity of assembly resources*.
- *Different cubatures* of the products and *interfaces* in the products significantly increase the *complexity of a potential automation*.

6.3.2 Limits of Design-for-Takt

The overall objective of MMAD and PD for MMA is to create the prerequisites for manufacturing a wide variety of products with a wide variety of options cost-efficiently in a flow assembly line. Concentrating on only one (cost) target is not expedient, as this can result in major disadvantages for other targets. For this reason, we set the following limits in our Design-for-Takt approach:

- One of the largest drivers of total costs is parts costs (Wiendahl et al., 2004; Boothroyd, 2005). For this reason, no Design-for-Takt measure should negatively affect parts costs.
- Further, it can be deduced from the DFA approach that the degree of standardization must not be reduced and the number of different parts must not be increased (Boothroyd, 2005).
- No measures should increase the assembly time of individual work operations, even if this minimizes utilization losses (Kesselring, 2021). Reducing the total assembly time is always preferable to minimizing utilization losses.
- Assembly quality should be given a higher priority in management attention; it is a sustainable, economic success factor. Unfortunately, the effects of decreasing quality are often not directly quantifiable, yet we set the condition that no measure from the Design-for-Takt should reduce assembly quality.
- The variance presented externally to the customer is a clear competitive advantage and great lever for the margin of products. For this reason, no Design-for-Takt measure may reduce the variance perceived by customers (Kesselring, 2021).

6.3.3 The Principles of Design-for-Takt

In the following, we present principles of Design-for-Takt that are intended to preemptively reduce an assembly's utilization losses, modification costs, and capital expenditures. To be consistent with the focus of our book, we always relate the principles to variable takt. Because Design-for-Takt is a preventive approach, variable takt does not support Design-for-Takt—it is the exact opposite. The Design-for-Takt supports and forms the basis for many of the concepts of AD for MMA (compare Sect. 1.4). At the end of this section, we summarize all the principles again in Fig. 6.10.

6.3.3.1 Targeted Flexibilization of Precedence Relationships

Precedence relationships are one, if not the, determining factor in line balancing of an assembly line (Kilbridge & Wester, 1962). No factor determines more strongly the degrees of freedom of a line balancing and thus the solution space in AD for MMA (Sect. 1.4). We have briefly described the mode of operation, background, and limitations in the practical use of a precedence graph in Sect. 2.6.4.

Precedence relationships can result from:

1. Technical restrictions of the assembly system or
2. Are constructively anchored in the design of the product

Ultimately, both restrictions create constraints on the sequence of the assembly process (Kilbridge & Wester, 1962). Technical constraints in the assembly system arise in the planning phase of assembly and are thus also changeable in this phase, usually with the use of investments and costs. Design precedence relationships arise in the product creation phase and can therefore also only be changed subsequently by changing the design of the product, usually in the context of a subsequent product generation or model upgrade.

Despite the fundamental importance of precedence relationships for line balancing, these are not explicitly documented in many companies and are usually only implicitly anchored in the knowledge of the employees, and again not completely. Possible reasons are described in Sect. 2.6.4.

In his work, Swist (2014) investigated the relationships between variant precedence relationships as well as the density of these precedence relationships and utilization losses. His main findings were (Kesselring, 2021):

- The specific order in which assembly operations must be performed can negatively affect balancing efficiency.
- There is a causal relationship between the density of precedence relationships and utilization losses. The higher the density, the higher the losses.
- Precedence relationships limit the achievable maximum of balancing efficiency (cf. Fig. 6.1, boundary line of the solution space).
- Precedence relationships limit the even distribution of assembly operations with large operation times.
- Precedence relationships limit the equal or targeted distribution/shifting of variance content.

Benefit

With this knowledge and the awareness that precedence relationships decisively determine the degree of freedom of the line balance, the density of precedence relationships can significantly influence the occurrence of utilization losses. If a defined balancing efficiency has to be achieved with a new or modified line balancing, the density of the precedence relationships influences the duration and/or the effort, and thus the modification costs, of this line balancing. A low density of precedence relationships should always be aimed for.

Precedence relationships are a great enabler for many methods and concepts to master variance in AD for MMA, they enable many Design-for-Takt approaches from Fig. 6.10. For example, variance can only be selectively shifted to special task workers or to pre-assembly if the precedence relationships allow it. The same condition applies to parallel work. Variable takt requires an even distribution of the workload of the products to be assembled along an assembly line, and low precedence relationships promote this.

However, precedence relationships do not only have a decisive influence on running costs (utilization losses) or on the costs of changing a line balancing. They also have a significant influence on investments. For example, effective Variant Management leads to as many common parts and common interfaces as possible and thus to standardized equipment and assembly systems. However, if these cannot be combined at a local point on the assembly line due to precedence relationships, operating resources and equipment must be installed several times. In Sect. 6.3.3.4, we describe the effect of having to invest in the expansion/extension of an assembly line if the precedence relationships force exclusively serial working on the product. If assembly activities can also be carried out in parallel by relaxing the precedence relationships, the assembly line can be shortened, which results overall in lower investments.

Now What? Make All Precedence Relationships More Flexible?

To dissolve all precedence relationships of a product and to strive for complete flexibility in the assembly sequence is technically and economically unrealistic. These efforts are disproportionate to the potential benefits (Swist, 2014). Research and concepts to make precedence relationships fundamentally and systematically more flexible have not yet been widely disseminated. There is a need to catch up here in the development of optimization methods!

In Design-for-Takt, flexible precedence relationships enable a number of other concepts to reduce utilization losses, modification costs, and investments in assemblies. We present these in the following sections. Limited resources (mostly design budgets) to reduce the density of precedence relationships should be targeted where they enable or support the following concepts.

6.3.3.2 Dealing with Variance, Additive and Alternative Variants

In his work, Swist (2014) demonstrates that one of the biggest drivers of utilization losses is additive variants. The reason for this is obvious, the operation time of the relevant sections is either zero or has its individual value ot_l (Sect. 2.6.8), i.e., the takt spread becomes "binary." Alternative variants behave very similarly to additive variants, although not with as many proven effective relationships (see Fig. 6.2). Theoretically, each additive variant could also be considered as an alternative variant, where the operation time (Sect. 2.6.8) of one specific alternative is zero. For this reason, exactly the same proposed solutions apply to the Design-for-Takt measures for alternative variants as for additive variants. Figure 6.2 visualizes the effective relationships between additive and alternative product tree items (BOM items) and utilization losses as demonstrated by Swist (2014).

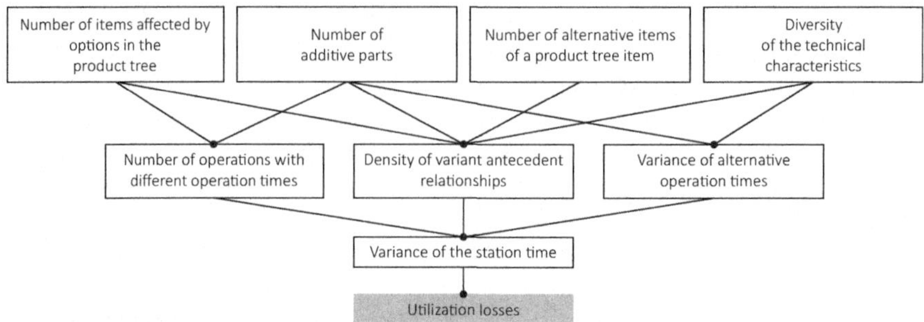

Fig. 6.2 Proven (Swist, 2014) and visualized (Kesselring, 2021) impact relationships that lead to utilization losses

The simplest way to minimize the resulting utilization losses would be to abandon additive variants, to reduce alternative variants, or to reduce the number of characteristics of both types of variants. However, this would lead to a reduction in external variance, which we have ruled out for Design-for-Takt measures.

The number of additive variants could be reduced via packages (Sect. 6.2)—the combination of certain additive variants is prevented. This also leads to a reduction of the external variance and is therefore not part of our proposed solution.

- The DFA method (Sect. 6.1) can be used to reduce assembly times in a targeted manner. Resources for new product development or modification are always limited; if these are used specifically to reduce the *assembly times of additive variants,* utilization losses can be reduced (Kesselring, 2021). The *targeted reduction of the takt time spreads of alternative variants* leads to the same positive result (Kesselring, 2021).

One note: If these limited resources can be used to reduce variant-free assembly times to the same extent, this is to be preferred! Because these reductions lead in direct proportion to a potential reduction of the takt time, and not, as with variants, with the proportion of their probability of occurrence (see Sect. 2.6.8).

- By using the modular principle, the interfaces of alternative variants can be specifically standardized (Swist, 2014). The takt time spread generated by these variants can be shifted to pre-assemblies, and the *standardized interface* minimizes takt time spread in the line (Fig. 6.3).
- One way of making the precedence relationships of additive variants more flexible is to structure the product tree (the BOM structure) in a targeted manner in product architecture planning (Swist, 2014). *Basic content should be separated from variance content*; this makes their precedence relationships more flexible and forms the basis for bundling variance in modules. Furthermore, *positively correlated variances* should be mapped *in separate product tree structures* and *negatively correlated variances* should be *bundled (Fig.* 6.3*).*

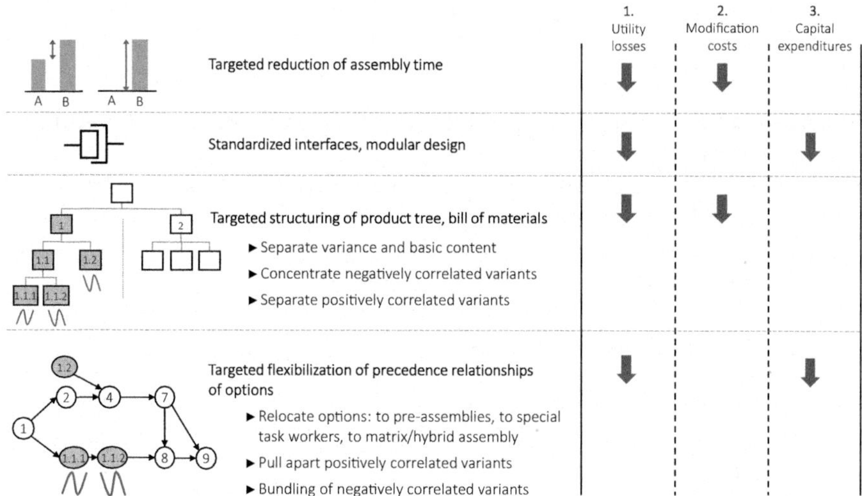

Fig. 6.3 Dealing with variance in Design-for-Takt (own representation adapted from Swist 2014 and Kesselring 2021)

- *Precedence relationships of option variants should be made flexible in a targeted manner* so that option variants can be integrated into modules (Swist, 2014), which are shifted from the assembly line. In this way, these option variants do not negatively influence the takt time spreads of the line. Flexibilization allows option variants to be bundled and moved to pre-assembly, to special task workers, to matrix assembly, or to hybrid stations. We describe an example of these applications with the Fendt assembly system in Sect. 7.1. In these areas, the local variance is decoupled and can be mastered more specifically outside the flow system (Fig. 6.3).
- If it is not possible to remove option variants from the line, the targeted flexibilization of their precedence relationships has a further advantage. Swist (2014) proves that the *targeted equalization of positively correlated variants* and the *concentration of negatively correlated variants* reduces utilization losses in the assembly line. This solution approach preserves the benefits that assembly in a clocked and flowing assembly line generates (Sect. 2.1). In order to increase the joint probability of occurrence of negatively correlated option variants, packages in which the combination of these option variants is promoted would be one way of specifically influencing the demand side (Fig. 6.3).

... and with variable takt?

- Additive variants always lead to local takt time spreads. Alternative variants lead to local takt time spreads if their operation times ot_l differ. For these reasons, variable takt in line balancing is not a suitable element for mastering this variance. However, if a customer-specific order contains a particularly large number of additive options or alternative variants with larger operation times, the takt time of these individual orders could be increased by an order-specific offset within the framework of workload-oriented series-sequencing, as we describe in Sects. 3.2.3 and 8.6.4, in order to minimize model-mix losses.

6.3.3.3 Aiming for Small Operation Times, Equalizing Large Operation Times

For the following Design-for-Takt approach, we resort to the definition of balancing efficiency from Sect. 2.6.11:

$$E_{AL} = \frac{\sum_{n=1}^{N} R_n}{N \cdot T_j} \cdot 100[\%]$$

with

E_{AL} Balancing efficiency of the assembly line
N Total number of stations (assumption: one worker per station)
R_n Weighted station time at station n
T_j Takt time of order j

The balancing efficiency describes the quality of the balancing process for the entire assembly line, from which the balancing efficiency for a station (we assume one worker per station or workstation) is derived as follows:

$$E_{WSn} = \frac{R_n}{T_j} \cdot 100[\%] = \frac{\sum_{l \epsilon L_n} (ot_l * P_l)}{T_j} \cdot 100[\%]$$

with

E_{WSn} Balancing efficiency at workstation n
L_n Index set of assembly operations l assigned to station n
ot_l Operation time of assembly operation l
P_l Probability of occurrence of assembly operation l

Takt losses, and thus utilization losses, occur with a balancing efficiency of less than 100%. The numerator of the formula for calculating the balancing efficiency is the sum of the weighted operation times (ot_l*P_l). For the line balancing it follows:

Firstly, if the smallest weighted operation time is, for example, 5% of the takt time T_j, the balancing efficiency will, in the best case, be between 95% and 100% at this workstation. The larger the smallest operation time, the greater the spread and the more likely lower balancing efficiencies are and thus greater takt losses. In short, the largest possible proportion of operation times should have the smallest possible ot_l in relation to T_j.

Secondly: Particularly large operation times components (with a high P_l) in relation to the takt time T_j significantly reduce the degrees of freedom in the balancing, because they make it less likely to completely fill the numerator of the

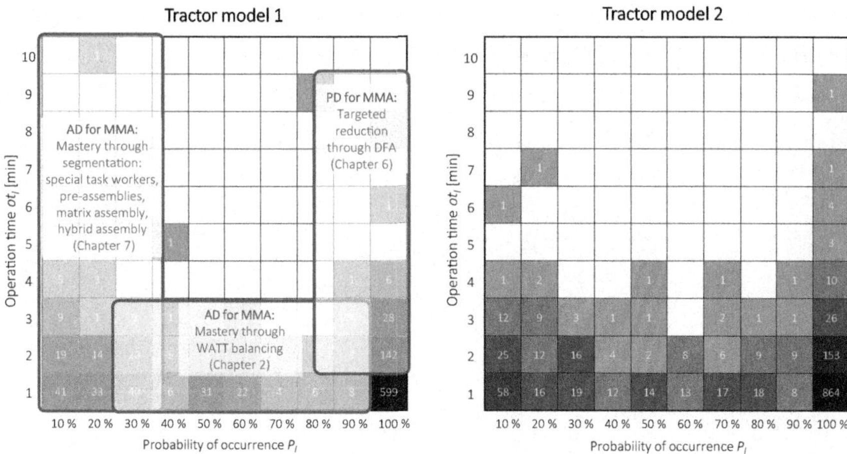

Fig. 6.4 Distribution of operation times for the Fendt tractor assembly case study (adapted from Kesselring, 2021) and elements of the Mixed-Model Assembly Design (MMAD)

formula for calculating the balancing efficiency. Or, to put it another way: the WATT balancing method presented in Sect. 2.6 leads to better balancing efficiencies with many small operation times (in relation to the takt time) with less effort. In short: operation times with large ot_l, in relation to T_j, should be avoided.

Consequently, it must be a Design-for-Takt approach that avoids non-divisible assembly operations that lead to large operation times in relation to the takt time. Based on practical experience, interviews with planners and team leaders in assembly and with the help of a case study (Kesselring, 2021), all time modules that exceed 11% of the takt time can be considered problematic (see the following case study Figs. 6.4 and 6.5). These not only lead to a lower balancing efficiency but also significantly reduce the degrees of freedom in line balancing and thus increase the effort required when making modifications to the line balance.

The initially simplest solution to avoid large operation times would be to divide the underlying assembly parts into several components (Kesselring, 2021). However, this violates one of the principles of Design-for-Takt from Sect. 6.3.2 and is therefore not considered.

- As described in the basics of DFA (Sect. 6.1), an assembly process consists of three basic activities: handling, joining, and fastening. In the classical design, it is often necessary for a worker to perform all three steps in a coherent sequence. A design that allows the *processual separation of the activities or separations within these activities* is preferable. For example, components should already be connected to each other after joining (e.g., by clamping) so that fastening can be performed separately.
- *Interrelated assembly activities*, where *several parts have to be aligned and connected at the same time*, must be *avoided*. If, for example, several hydraulic

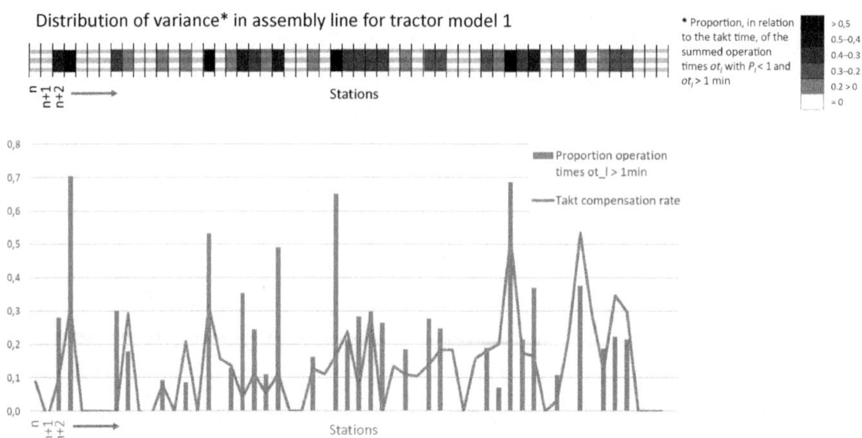

Fig. 6.5 Distribution of operation times $ot_l > 1$ min and takt compensation rate per station

pipes are connected to each other via T-pieces or other manifolds, this must be carried out coherently by one worker in order to avoid tensions between the pipes due to simultaneous alignment. Such forced couplings of work operations are to be avoided.

- If *large and interconnected assembly operations (composite assemblies)* cannot be avoided, they should be *located as evenly as possible in the precedence relationships or designed as independently* as possible. In this way, they can be distributed evenly over the assembly line. In the following case study, we show how grouping these large operation times have a negative impact on the balancing efficiency.

Case study final assembly Fendt Tractor

The importance and impact of large operation times are illustrated by a case study from Fendt tractor assembly. All of the following data are from this case study (Kesselring, 2021). In the data, two tractor models are considered in one section of tractor assembly, the variable takt time of the two models differs and is between 7 and 9 min. For each model, over 1000 work operations are in the scope of consideration. More detailed specifications cannot be given, but more details are also not necessary to clarify the meaning and effect of the operation times, their distribution and probability of occurrence.

Figure 6.4 shows that the proportion of operation times over 1 min is comparable for both models: 23.5% for model 1 and 21.2% for model 2. First of all, it can be seen that by far the largest proportion of operation times are under 1 min and here the focus is on the probability of occurrence of 100% (basic contents). Noticeable in both types is a concentration of larger operation times with low P_l (left side in both diagrams), this can be explained by the targeted use of special task workers for these work operations in our case study. Background information on this and further possibilities to master larger operation times with a low probability of occurrence P_l are explained in Sect. 7.1 of the Fendt assembly system.

For our considerations in this chapter, it should be noted that larger operation times with a low probability of occurrence are less significant for the weighted station time R_n because they are only included with their weight in the numerator of the formula for calculating the balancing efficiency. These effects can be mastered in the WATT balancing method (Sect. 2.6.8) by means of overtakting and open station boundaries (Sect. 2.6.10). The lower middle region of the matrix in Fig. 6.4 is particularly suitable for these solution strategies. Problematic for the balancing efficiency, and thus takt losses, are primarily larger operations with a high P_l. The limiting resources in the design phase of a product or later improvement initiatives should concentrate precisely on this area with a DFA approach.

To analyze the effects of large operation times, we examined the correlation between the proportion of operation times greater than 1 min per station and takt compensation in our case study from Fig. 6.4 (Sect. 2.6.11).

To our surprise, we could not find any correlation in the first step for both tractor models. However, if we excluded all basic contents (basic work operations; $P_l = 1$) from the calculations and only examined the correlation between the summed operation times ot_l greater than 1 min with a probability of occurrence $P_l < 1$ and the takt compensation rate, we obtained clearly significant correlations for tractor model 1. In other words, an increasing density of variant times greater than 1 min (corresponding to approximately 11% of the takt time in our case study) correlates positively with the takt compensation rate. The correlation for tractor model 1 is 0.41. The statistical p-value of the correlations is 0.02 (this corresponds to a significance level of at least 95%), the correlations can thus be considered significant. We can visually represent this correlation in Fig. 6.5.

The takt compensation is the reciprocal of the balancing efficiency (Sect. 2.6.11). The proven correlations in the visualization in Fig. 6.5 clearly shows that a high density of variant times ($P_l < 1$) greater than 1 min ($ot_l > 1$ min) and the balancing efficiency of the stations correlate negatively with each other.

… and with variable takt?

- Variable takt allows products with a greater total assembly workload to be run at a greater takt time in the assembly line. We have described how this works in Chaps. 4 and 5. If a variable takt is used, it also allows larger individual operation times in the same ratio for these products. Since larger operation times are local events, in these cases variable takt can only provide limited support in reducing utilization losses.

This section also shows that the Design-for-Takt approaches (Fig. 6.6) cannot be directly supported by variable takt. The causal relationship goes in the other direction. It is becoming increasingly apparent that variable takt is supported or enabled in the first place by Design-for-Takt.

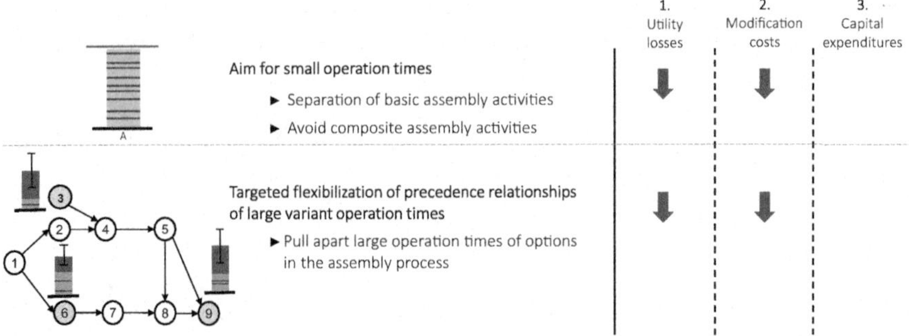

Fig. 6.6 Dealing with large operation times in Design-for-Takt (own illustration adapted from Swist 2014 and Kesselring 2021)

6.3.3.4 The Ideal Flow Assembly: Even Distribution of the Workload Around the Cubature of Assembly Objects

Our final principle of Design-for-Takt is specifically aimed at reducing investments in assembly. In the previous sections, we have already shown that by standardizing the interfaces and making the precedence relationships more flexible, operating resources and equipment can be grouped and standardized in order to reduce their procurement costs. Another key driver for investments in an assembly line is the space required for the assembly line. The more compact, and therefore shorter, the assembly line can be, the less needs to be invested in (building) infrastructure. In existing structures (brown field), additional space for a necessary assembly expansion is very often no longer feasible, even with increased budgets.

In Fig. 6.7, we show in simplified form the cause-and-effect relationships that essentially determine the length of an assembly line. An area is the product of length and width. For simplicity, we will concentrate on the length factor. We will show that the extension of the assembly line can be influenced by our principle, a levelled distributed workload along the cubature of the products. The length of the entire assembly line is the product of the station length and the number of stations in the assembly line (see Fig. 6.7). According to our derivations in Chap. 2, the number of stations required results from the quotient of the total assembly workload to be represented in the line and the product of takt time, worker density (workers per station), and balancing efficiency E_{AL} (Sect. 2.6.11).

From the assembly point of view, the influencing factors can be evaluated as follows:

Volume	*External specification*: Can be regarded as an external variable of the market. Assemblies should be able to follow fluctuations in demand quantity as flexibly as possible in order to avoid inventories, but also supply bottlenecks.
Assembly line runtime	*AD for MMA*: Can be designed by management. Volume changes should first be responded to by adjusting line run time through changes in work schedules or shift systems.

(continued)

Total assembly workload in the assembly line	*External specification*: On the one hand, the total workload required to assemble the product is determined by the product itself and the properties depicted, and is thus specified by the market and the customers. *PD for MMA*: The methods of DFA and Variant Management can be used to specifically reduce assembly workloads. *AD for MMA*: Segmentation can be used to selectively move assembly content to pre-assembly or other assembly areas, i.e., workload is removed from the assembly line. In Sect. 2.1 we describe why we are convinced that a large part of the assembly workload should remain in the assembly line.
Worker density	*PD for MMA*: The cubature of the product itself, but above all the distribution of the workload on or along this cubature, determines how many workers can work on the product in parallel. The more uniform the distribution, the greater the workforce density can be. The workforce density (wd) indicates how many workers are employed on average per station: $wd = \frac{W}{N}$.
Balancing efficiency (E_{AL})	*AD for MMA*: The theoretical balancing efficiency to be achieved is set in the line balancing. Here, variable takt can make a major contribution, because, for products with a smaller total assembly workload, the takt time is reduced, and thus the balancing efficiency is increased. It could also be said that the available line length is utilized more efficiently; we describe this effect in Sect. 4.4.4.
Station length	*External specification*: The product itself essentially determines the extent of the individual station. Further factors can be the selected transport system or safety requirements.

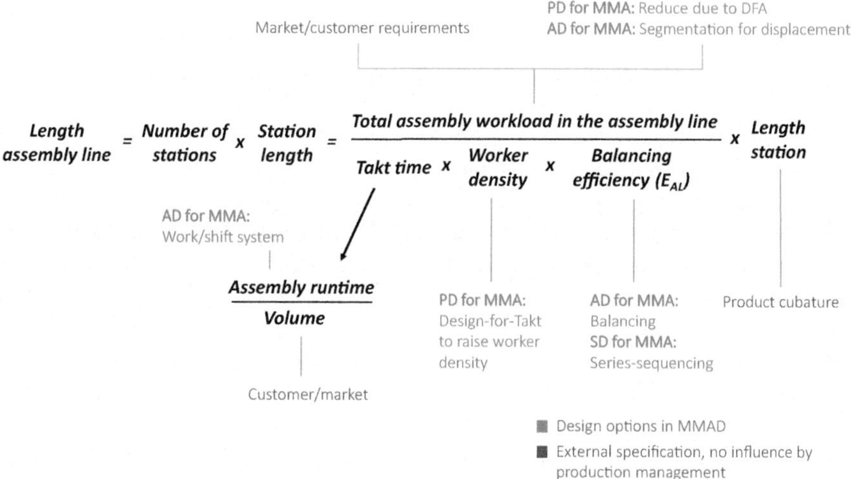

Fig. 6.7 Simplified representation of the cause-and-effect relationships for the expansion of an assembly line

Comparison of the influencing factors Figure 6.7 shows that the assembly workload correlates positively with the length of the assembly line. As already described several times, we do not want to reduce the assembly workload in the line in principle. If the factors assembly running time, worker density and balancing efficiency are increased, this reduces the necessary length of the assembly line—these factors are in the denominator of the equation in Fig. 6.7. The balancing efficiency E_{AL} can reach a maximum value of 1 and in classic assemblies lies between 0.7 and 1, its leverage on the length of the assembly line is thus limited. Assembly running time can be doubled when switching from a one-shift to a two-shift system, which represents a much greater leverage. Worker density represents a similarly large, if not larger, lever; its reciprocal is directly proportional to the length of the assembly line. In short: The more workers can be deployed in one station at the assembly object, the shorter, in direct negative proportionality, the assembly line can be designed. Investment costs can be significantly reduced. Existing assembly lines are often faced with the challenge of having to handle an increased total volume or products with a greater overall assembly workload. If the number of stations is given (brown field) and the assembly system has already been converted to a multi-shift system, the worker density is the only way to compensate, following the equation in Fig. 6.7.

In Fig. 6.8 we present a static picture of an assembly station with a possible staffing of 6 workers $wd = 6$ at an assembly object with a length of 5 m. From this simplified representation, it quickly becomes clear that such a high worker density is only possible if the workload is equally distributed along the cubature of the product. If the main part of the assembly effort were located in the front part of the product, workers W4, W5 and W6 could not be loaded to the same extent as workers W1, W2, and W3. More stations would be needed to handle the entire workload.

The static representation of the worker density at a station in Fig. 6.8 does represent the necessity of an evenly distributed workload along the cubature of the assembly object. The static form of representation is used in most publications, but it only falsifies the real processes in a flowing assembly, because the assembly object moves on continuously. Following the theory of the worker triangle, the worker flows with the assembly object from the start of the assembly station, at the end of the station, after finishing his/her activity, he/she collects information and material for the next job and returns to the start of the station (Sect. 2.3.1). Walking is waste (Sect. 3.1) and should be avoided. The sides in an ideal worker triangle would thus have zero length—resulting in an ideal assembly point where the assembly objects flow past him/her and the worker does not have to leave his/her position. For this reason, the work operations in the SWS of an assembly object are arranged in such a way that they are located in the front part of the assembly object at the beginning of the worker's activity. As the takt time progresses, the worker performs activities along the assembly object; towards the end of the takt time, he/she should perform activities at the rear part of the assembly object.

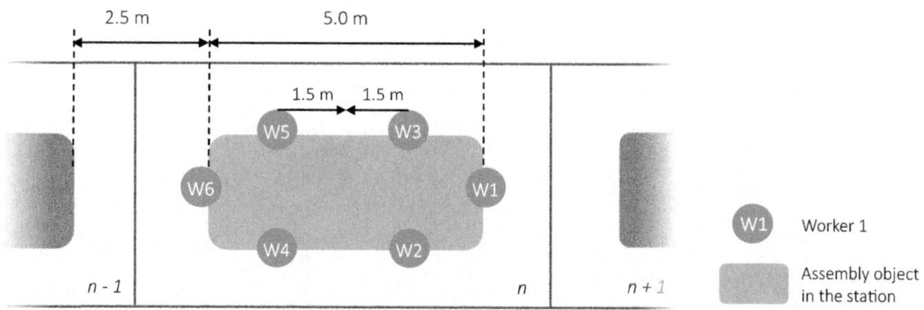

Fig. 6.8 Static representation of the worker density at an assembly station

Definition of assembly access surface If an assembly object is moved past a stationary worker (no change in position), the assembly access surface results from the surfaces on the assembly object that can be reached by the worker. For most assemblies, this results in the assembly access surfaces of the left and right sides. It can also be the bottom of the object, if the object is guided above the positioned workers.

Following this model of continuous movement of objects, the assembly workload should be equally distributed per unit length of the assembly access surface. In this way, walking distances, and thus waste, can be reduced and the individual workers can be positioned at minimum distances on the assembly line. If the workload is completely equally distributed and the assembly objects can be placed at zero distance on the assembly line, this results in the theoretical minimum length of the assembly line. A station could thus no longer be defined by the cubature (the length) of the assembly object, but by the minimum radius of movement of a worker.

From Fig. 6.9, *there are two variables that describe the distribution of workload on an assembly object:*

wlo*: *workload per unit length on the assembly object
wls*: *workload per unit length of assembly access surface

To minimize the necessary movements of the workers and to maximize worker density, these variables should have the same value within each product and across all products of an assembly line.

For these reasons, it must be an objective of Design-for-Takt to distribute assembly content equally around the cubature of the product in the design phase. The defined variables *wlo* and *wls* can serve as target variables in the design phase to limit the investments of an assembly.

... and with variable takt?

- Following the ideal line occupation from Fig. 6.9, no variable takt would be possible here. The distances between the assembly objects can no longer be changed. A variable takt would also no longer be necessary! The ingenious

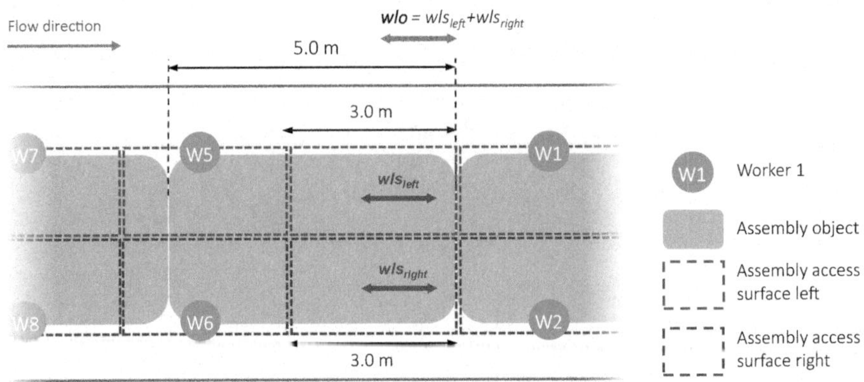

Fig. 6.9 Ideal worker density using the example of an assembly object with a length of 5 m and a worker radius of 1.5 m

thing about the variable takt is that it compensates for an existing inequality in the distribution of the workload along the cubature (different *wlo*) of the assembly objects. In other words: If two assembly objects with different total assembly workload are assembled in a flowing assembly line in a fixed takt, the distribution of the workload per unit length, the *wlo*, will be different. If, for example, the variable takt reduces the distance between the products with a lower total assembly workload, this compensates for the different *wlo* values.

6.3.3.5 Summary of Design-for-Takt Principles

With the concepts of Design-for-Takt presented here (Fig. 6.10), we do not claim to be exhaustive. On the contrary, we would like to encourage further research into this topic area and to implement it even more strongly in practical day-to-day business.

In Mixed-Model Assembly Design (MMAD), Design-for-Takt is an element of Product Design for Mixed-Model Assembly (PD for MMA), which has a strong preventive effect. If a product is already defined, and the BOM and 3D data sets are created, it becomes very difficult to implement the concepts shown here. For this reason, the elements of PD for MMA should be strategically implemented in product development. **Design-for-Takt can significantly increase the solution space of later line balancings and thus contribute to a sustainable reduction of utilization losses, as well as reduce the effort of new line balancings.** Capital expenditures are reduced through standardized interfaces, flexible precedence relationships, and even distribution of workload along assembly access surface. The implementation and success of variable takt and VarioTakt can be significantly supported by the concepts of Design-for-Takt. In many cases, it can be assumed that Design-for-Takt enables the use of variable takt in the first place.

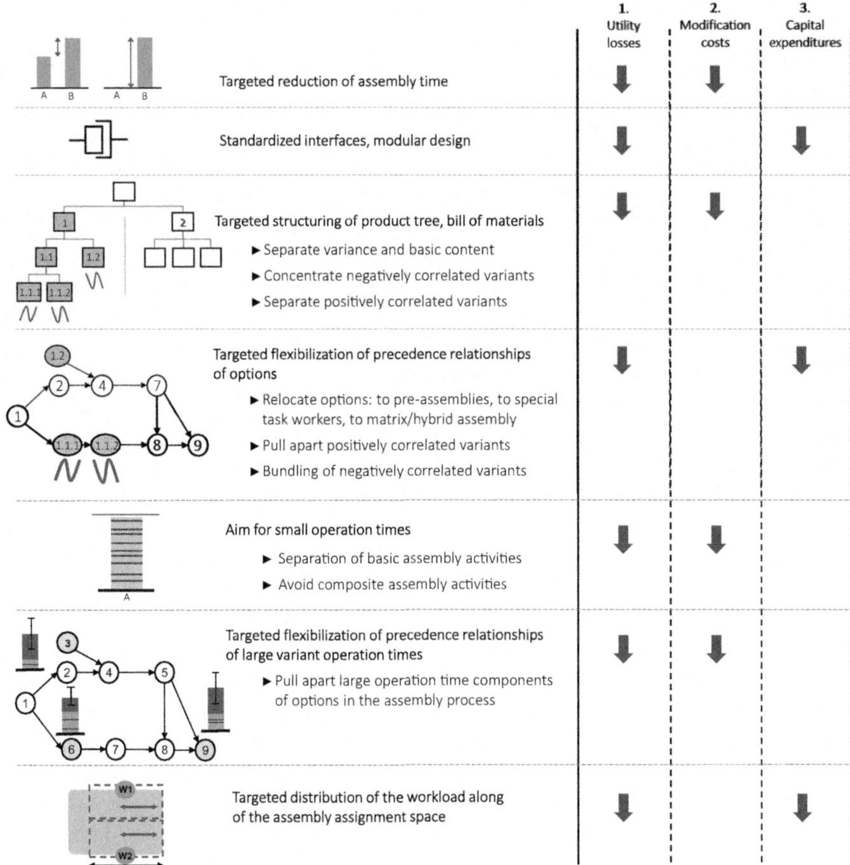

Fig. 6.10 Summary of Design-for-Takt methods

References

Andreasen, M. M., Kähler, S., & L'Und, T. (1988). *Design for assembly* (211 p.). Springer.

Boothroyd, G. (2005). *Assembly automation and product design* (536 p.). Taylor & Francis. https://doi.org/10.1201/9781420027358

Kesselring, M. (2021). Product design for mixed-model assembly lines. Master Thesis, WHU, Otto Beisheim School of Management.

Kilbridge, M., & Wester, L. (1962). A review of analytical system of line balancing. *Operations Research, 10*(5), 591–742. https://doi.org/10.1287/opre.10.5.626

Rapp, T. (1999). *Produktstrukturierung – Komplexitätsmanagementdurch modulare Produktstrukturen und –plattformen* (172 p.). Gabler Edition Wissenschaft.

Swist, M. (2014). *Taktverlustprävention in der integrierten Produkt- und Prozessplanung* (328 p.). Apprimus Verlag.

Wiendahl, H.-P., Gerst, D., & Keunecke, L. (2004). Variantenbeherrschung in der Montage. *Springer Verlag*. https://doi.org/10.1007/978-3-642-18947-0

7 Mastering Variance in Assemblies: The Fendt Assembly System and Matrix Assembly

The increased demands on mixed-model assemblies, especially of type 3, have not only led to the development of the VarioTakt at Fendt, but numerous other concepts in the field of AD for MMA (Sect. 1.4) have emerged in the industry to master variance in mixed-model assemblies. Matrix assembly is currently one of the most discussed concepts in the field of AD for MMA and the one that has received a lot of attention in recent research work.

One of the pioneers of matrix assembly in the automotive environment is Audi. The aim of using matrix assembly is to break away from the constraints of line assembly. The fact that this is not always necessary and can sometimes be counter-productive is described in Sects. 7.2 and 8.1.

We see the greatest future chances of success in a combination of an assembly line with matrix assembly, as we classify them in Sect. 2.2 as hybrid assembly structures. Here, the matrix can be positioned before or after the line assembly or interrupt the line assembly at several points. The simplest variant, a M1 matrix, a punctual shear-off, is presented in the following Sect. 7.1 in the "Fendt assembly system" and in Sect. 8.2 "hybrid assembly."

Figure 7.1 shows possible combinations of line and matrix assemblies. The LM structure (*L*ine-*M*atrix structure) is the easiest structure to implement in practice, since changes to the sequence in the matrix, triggered by disruptions or different process times, have no effect on the already completed line assembly. LM structures can unfold their advantage when option variants can be bundled at the end of the assembly process—local bundling (Sect. 2.6.11).

If we assume that the matrices in Fig. 7.1 consist of three independent cells, we refer to these matrices as an M3 matrix. In addition, in our example, each of the three horizontal cells is identically duplicated three times. An order can thus pass through none, one, two, or three of the cells, but cannot visit any of the identical cells twice. Each of the three (horizontal) matrix cells is different in its structure and the content of work that can be processed in it. In general, an MX matrix consists of X cells in the direction of product flow with one or more parallel (i.e., duplicated) cells. Products can then be controlled in any way through these X process steps. We

P. Bebersdorf, A. Huchzermeier, *Variable Takt Principle*, Management for Professionals, https://doi.org/10.1007/978-3-030-87170-3_7

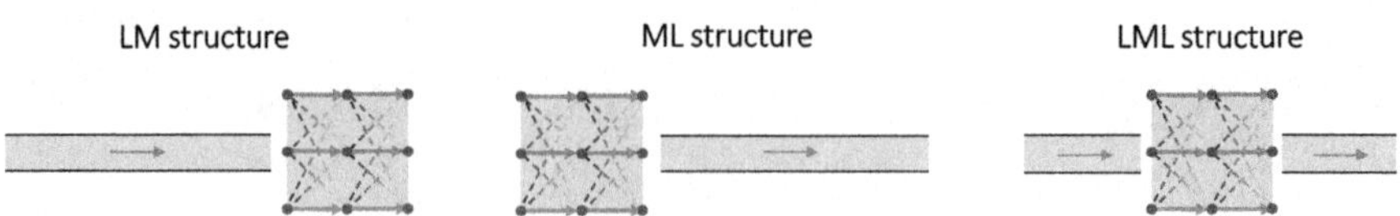

Fig. 7.1 Structural forms of hybrid assemblies with an M3 matrix

describe a very simple application of an M1 structure in the following Sect. 7.1, using Fendt's tractor production as an example. A M1 matrix exists when there is only a single matrix cell or when there are multiple identical cells (duplicates in parallel), where each order can only passes through one (or none) of these cells.

LML structures (including multiple LML structures) offer the greatest potential in dealing with variance, especially in the case of local technological modifications such as e-drives or autonomous driving, because here large local takt time spreads can be accommodated during the assembly process. In Sect. 8.2, we show how BMW is working with RWTH Aachen University on a concept for the practical implementation of a hybrid assembly (Kampker et al., 2021).

7.1 The Fendt Assembly System

With the Fendt assembly system, we would like to present a successful combination of the most diverse elements from AD for MMA to master variance in flow assemblies.

A matrix, albeit greatly simplified, is also used by Fendt, the pioneer of variable takt, in its tractor production. It is a M1 matrix with five parallel identical cells. Thus, the tractors pass through a maximum of one of these cells, each cell is identically equipped, and the same activities can be performed in each cell. To manage the complexity of tractor production, Fendt's assembly system essentially consists of four main elements (Fig 7.2):

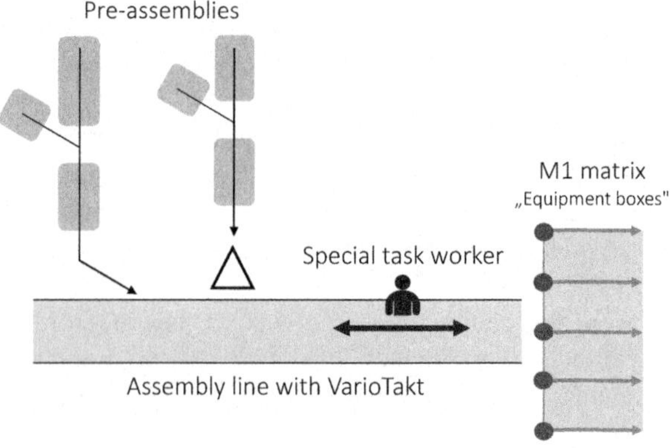

Fig. 7.2 The Fendt assembly system

- Pre-assemblies
- Special task workers
- Line assembly in VarioTakt
- M1 matrix

The smart combination of these elements makes it possible to manage assembly workload differences of up to 60% between the tractor series and additional option variance within the series (see Sect. 3.2.5).

7.1.1 Flow Assembly in the VarioTakt

Due to the numerous advantages of production in a continuous flow assembly line (see Sect. 2.1), Fendt's objective is to map as much assembly content as possible in the assembly line. The VarioTakt is the most important enabler to be able to efficiently assemble eight different products in a total-mix in a single line. In Chaps. 2 and 4, we have described in detail the principles and mode of operation of variable takt and the VarioTakt. If the takt time of the 700 Vario series is set to 100% as a reference, the smallest series, the 200V Vario, is assembled with 91% takt time and the 1000 Vario with 126% takt time (Fig. 7.5). We only use relative values here and in the entire Sect. 7.1 in order not to publish competition-relevant data, but we can still present an overview of the mode of operation of the Fendt assembly system and variable takt.

7.1.2 Pre-assembly

Pre-assembly lines can be directly linked to the assembly line in a clocked manner or they can operate without balancing into a buffer (Fig. 7.2). They always work order-related and thus not order-neutral in batch sizes.

Pre-assembly mainly fulfills four tasks:

1. Assembly of modules that cannot be assembled in the assembly line due to technical restrictions
2. Assembly of option variance of all products in order to minimize selective utilization peaks in the line → Avoidance of local takt time spreads (Sect. 2.6.11); the classic task of pre-assembly
3. Assembly of basic content with large total assembly workload to reduce the spread of the VarioTakt → Reduction of global takt time spreads (Sect. 2.6.11)
4. Workstations to integrate performance-impaired employees who cannot work in a flow assembly line

If space permits, pre-assembly should be placed as close as possible to the assembly station of the main line. This minimizes waste in the form of transport or inventory. With increasing automation of material transport, this restriction could play a lesser role in the future. If pre-assembly is assigned to the same organizational unit as the assembly station, there are further advantages of a value stream-oriented organization, such as easier defect traceability within one unit of responsibility or continuous improvement of the entire value stream. Capacity needs are calculated for each shift and pre-assembly staffing is adjusted accordingly. The expansion of smaller and especially variant-rich type 3 mixed-model assemblies with increasing product and option variance, leads to overcrowded material staging areas directly at the assembly line. An often-used solution is the switchover of material supply traveling in kitting carts which are attached to the conveyor system. In these cases, it would be possible to place the pre-assemblies in the kitting zones of these carts and to supply the pre-assembled components with the kitting carts—automated if possible.

7.1.3 Special Task Workers

Special task workers are employees that are located in the assembly line but can be planned and viewed as pre-assemblies. They could thus be described as floating pre-assemblies within the assembly line. They are not comparable to flexible workers as we describe them in Sect. 2.3.2. Special task workers have permanently assigned assembly work that they can execute over a larger area (several stations) in the assembly line. Predominantly, special task workers receive larger variance workload with a low probability of occurrence P_l. This assignment of work operations makes it possible to significantly increase the degree of utilization of all other line assembly employees—or in short: the work operations that would massively worsen line balancing are taken over by the special task workers. The activities of the special task workers are not balanced, but validated daily via a capacity calculation and, if necessary, the capacity of the special task workers is adjusted—as in the case of pre-assembly. Real-time control of the special task workers is necessary, because they have to perform their activities at the right time at the right station. Common MES systems combined with wearables such as smartwatches solve these tasks. The use of special task workers should be kept to a minimum; on the one hand, they require a very high level of qualification but cancel out the classic advantages of line assembly (Sect. 2.1) since they have a poor efficiency (value added time to capacity installed). On the other hand, they enable significantly improved balancing efficiency of the entire line by taking over rare variants.

7.1.4 Matrix Assembly

At the end of its assembly line, Fendt uses a simplified matrix in which no cross flows are possible. To be more precise, this is the simplest form of a matrix structure with one level, a M1 matrix, and at Fendt it comes with five parallel cells. The average takt time of this M1 matrix corresponds therefore to five times the takt time of the assembly line. These are identical cells, and each tractor passes through exactly one of these cells with an order-specific, individual stationary takt time. This simplified structure means that real-time control is not necessary, so planning and scheduling this small matrix is still very simple. However, due to the lack of a real-time control, disturbances cannot be compensated. The allocation of tractors and workers is calculated once a day, and reactive adjustments are not provided. The control system can be easily implemented with standardized, very simple software solutions. The matrix is deliberately placed at the end of the line assembly, as the sequence changes only have to be compensated for in the final inspection. Since almost no material has to be added in this area and only a small number of employees (compared to the entire assembly line) are busy, the resulting sequence changes can be dealt with. In this matrix, which is called "equipment boxes," significant option variant-dependent components are assembled which i) cannot be assembled in the assembly line due to technical restrictions or ii) would significantly reduce the balancing efficiency of the entire assembly line due to their large takt time spreads. This way, local option variance is grouped. Examples would be front and wheel weights of up to 3.5 tons or front loader assembly. Due to the numerous possible combinations of these options, the stationary takt in the equipment boxes or matrix cells can vary between 20 and 190 min. We describe the advantages of using matrices in detail in the following Sects. 7.2 and 8.1.

7.1.5 Mastering Global and Local Variance with Elements of AD for MMA

In the Fendt assembly system, different elements of AD for MMA are combined to handle both global and local variance.

- *Special task workers* and *pre-assemblies* are used to compensate for local workload peaks by options, but also to cover a part of the global variance of products with a high total assembly workload.
- Local variance is mapped in *pre-assemblies* and, if possible, assembled with standardized interfaces in the assembly line. In addition, global variance, basic product content with a high total assembly workload, is deliberately relocated to pre-assemblies.
- Local variance is specifically compensated with the *WATT method in the assembly line*. The line is selectively enriched with local variance, i.e., options of products with low total assembly content slightly level the global variance (see Figs. 7.4 and 7.5).
- At the end of the assembly line, additional local variance and customer requirements that cannot be assembled in flow mode are mapped in a *M1 matrix*, the equipment boxes.
- By using *variable takt*, the global variance between products on the assembly line is mastered (Fig. 7.3).

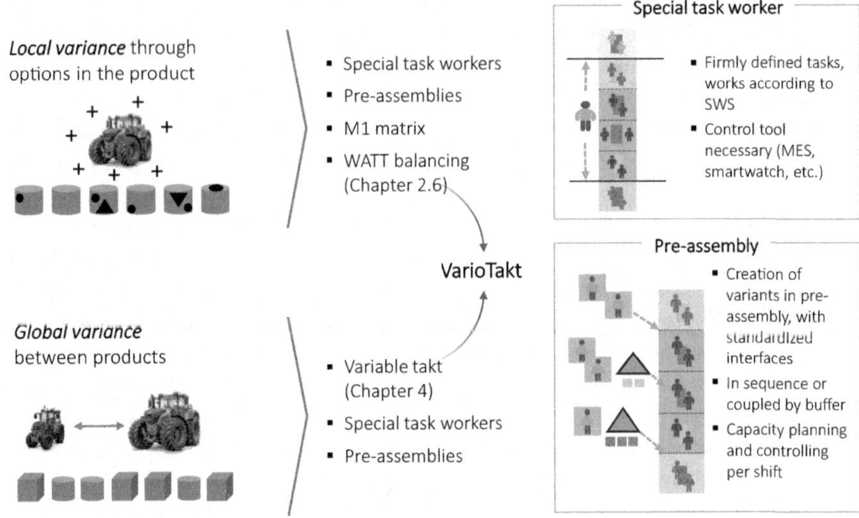

Fig. 7.3 Elements for mastering global and local variance in the Fendt assembly system

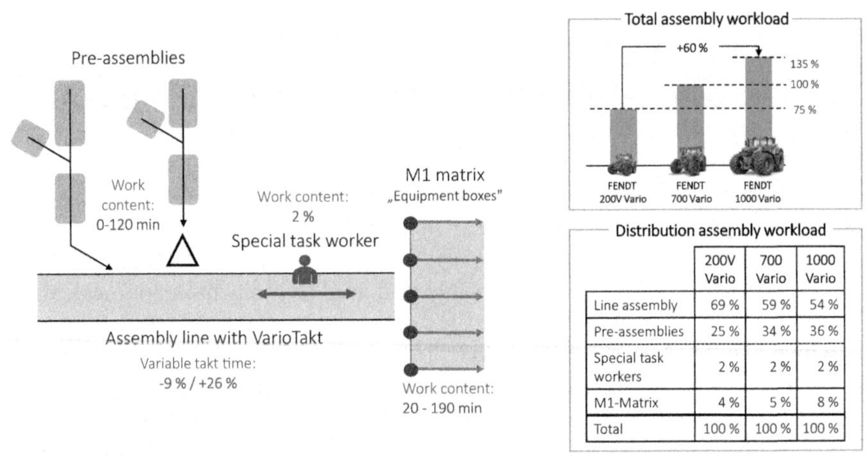

	200V Vario	700 Vario	1000 Vario
Line assembly	69 %	59 %	54 %
Pre-assemblies	25 %	34 %	36 %
Special task workers	2 %	2 %	2 %
M1-Matrix	4 %	5 %	8 %
Total	100 %	100 %	100 %

Fig. 7.4 Key figures in the Fendt assembly system

7.1.6 Effects of the VarioTakt in the Fendt Assembly System

Figure 7.4 shows the four structural elements of Fendt's tractor assembly, for quantification we use the 200V Vario, 700 Vario, and 1000 Vario types. In this system, several hundred tractors are assembled per week, divided into eight different tractor models, in a total-mix. The assembly line in Fig. 7.4 is interrupted by the paint shop, for simplicity, this has been excluded in the diagram. Most of the assembly workload for all types is shown in the main assembly line with a variable takt. The total mastered assembly workload in our case study has a spread of 75–135% (top right image in Fig. 7.4).

Pre-assemblies are arranged along the entire assembly line in a value stream-oriented manner; due to the increased use of attached traveling kitting carts in the future, these will increasingly move into the kitting zones. Some of the pre-assembly lines are connected to the main line in a clocked manner, while others are decoupled via buffers. Due to the assignment of activities to the pre-assemblies based on the four focal points from Sect. 7.1.2 and due to the variance of eight different tractor models, the assembly contents of the individual pre-assemblies vary from zero to 120 min. 25% of the assembly content is represented in the pre-assemblies for the smaller type 200V Vario, 36% for the 1000 Vario. If the difference in the total assembly time of both types is taken into account, the 1000 Vario has 2.6 times the assembly workload of the 200V Vario. The pre-assemblies are a crucial element in reflecting the variance of the total assembly workload. Without the variable takt of the main assembly line, however, pre-assemblies would have to represent an even larger part of the variance workload.

As described in Sect. 7.1.3, we consider special task workers as "floating pre-assemblies." Figure 7.4 shows that special task workers cover only a very small part of the total assembly workload, but this is crucial for a high balancing efficiency E_{AL} of the assembly line. Special task workers take on complex option variants with a low probability of occurrence. Due to the limitation of overtakting in the WATT (see Sect. 2.6.9) with $\beta = 15\%$, a significantly worse balancing efficiency would be the result without the use of special task workers.

The equipment boxes, which represent the matrix used in the Fendt assembly system, fulfill a similar task. Here, it can additionally be seen that significantly more content is mapped in the matrix for the 1000 Vario than for the other two types.

To emphasize it again at this point: It is the goal of the Fendt assembly system to keep the share of value added work within flow assembly as high as possible. We have described the motivation for this in detail in Sect. 2.1.

Despite the fluctuations in the total assembly workload between the tractor models of 75–135%, the relative share of the assembly workload of the 1000 Vario type in the assembly line can be kept high due to the VarioTakt (Fig. 7.4). Without this effect, with a fixed takt, the assembly line would have to be artificially enriched with even more pre-assembly content for smaller types (200V Vario) or even more assembly content would have to be shifted to pre-assembly for large types (1000 Vario). Or a significantly lower balancing efficiency is accepted, since the line balancing would then have to be oriented to the largest type, the 1000 Vario. The VarioTakt avoids this and does so without the compromise of low balancing efficiency. If we set the assembly content of the 700 Vario type in the assembly line to 100%, the VarioTakt will map 121% comparable assembly content in the assembly line for the 1000 Vario (Table 7.1 and Fig. 7.5)—this is almost two-thirds of the total difference in assembly workload! Due to the variable takt, the assembly line adapts to its load independently; we have described this effect in Sects. 4.4.2 and

Table 7.1 Relative assembly workload in the Fendt assembly system compared to the 700 Vario series

	200V Vario (%)	700 Vario (%)	1000 Vario (%)
Line assembly	**88**	**100**	**121**
Pre-assemblies	54	100	141
Special task workers	64	100	126
M1 matrix	53	100	207
Total	75	100	135

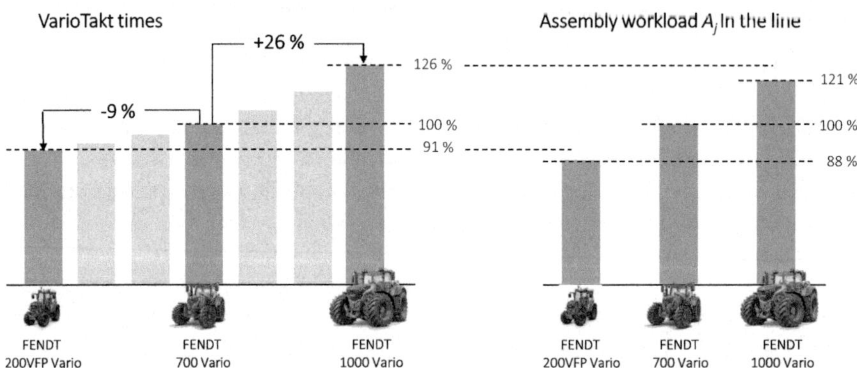

Fig. 7.5 VarioTakt—takt times compared to the coverage of the assembly workload in the line

4.5.4. Without the use of variable takt, these contents would have to be additionally covered by pre-assemblies, special task workers, or matrix assemblies—if this were technically possible. For the 200V Vario, it is only 88% assembly content in the assembly line compared to the 700 Vario.

Figure 7.5 shows very clearly that the spread of the takt times of approximately 35% is reflected in the spread of the assembly content of the line.

*A note from practice: The takt time of the 1000 Vario is 26% higher than the takt time of the 700 Vario type, yet in our case study only 21% more assembly content is mapped in the assembly line (*Fig. 7.5*). Why? Surely it should also be 26%. The answer to this question lies in the practical nature of production management. Resources, including those for continuous improvement, are limited—the 700 Vario is the volume model, and the 1000 Vario is the most powerful tractor, but not a volume model (see* Sect. 4.1.1*). Thus, the resources for optimizing the balancing and improving the line efficiency were concentrated on the type 700 Vario. The balancing efficiency of the 700 Vario is thus significantly higher than that of the 1000 Vario—hence the difference between takt time and coverage of the assembly workload in the line. A similar effect can be observed with the 200 V Vario, which should cover approximately 91% of the assembly content of the 700 Vario in the line, but due to the lower line efficiency, only 88% of assembly* *content is covered in the line. There is still significant potential for improvement in Fendt's assembly—just like in any other company.*

7.1.7 Conclusion Fendt Assembly System

Figure 7.4 is a highly simplified representation of the combination of different elements from AD for MMA in a LM structure, thus showing not only the basic structure of the Fendt tractor assembly but also a way how to integrate variable takt with matrix assemblies (of type M1) to many currently existing assemblies without major effort. The Fendt assembly system further shows the possibility to combine different elements from AD for MMA, in order to manage the global variance between the different products of the assembly, on the one hand, and to compensate the local variance of options, on the other hand.

We have described the prerequisites, the mode of operation, and the benefits of variable takt and VarioTakt in detail in the previous chapters of this book—so *nothing stands in the way of their application!* In the next section, we will go into more detail about matrix assembly and compare it to the VarioTakt.

7.2 Matrix Assembly: Or, How to Break Away from Clocked Operations?

Successful assemblies will have to master more variance in the future. Increasing customer individualization, disruptive new innovations or technology diversification and the hedging of global risks will lead to smaller and more variant-rich, perhaps already customer-specific assemblies (see also Sect. 1.2).

On the one hand, the (fixed) takt and thus continuous flow, is one of the basic elements of any assembly line and the basis of many managerial advantages (Sect. 2.1). On the other hand, a fixed takt time is the stumbling block of every production manager and assembly planner trying to master variance in the assembly line. Variable takt and the VarioTakt remove these obstacles. Despite the existence of this concept, which is still far too little used in practice, many production managers have been fascinated for some years by the idea of being able to free themselves from the burden of takt time and to be able to produce independently of a fixed takt. The idea of matrix assembly, also referred to as modular assembly or flexible cell manufacturing, is based on this desire. Matrix assembly has enjoyed increasing interest from manufacturers, consultants, and researchers in recent years (Kern et al., 2015; Greschke, 2016; Küpper et al., 2018; Bányai et al., 2019). In Sect. 2.2, we classified this assembly structure alongside other organizational forms of production.

Since our work is not intended to focus on matrix assembly, we will only briefly describe the most essential prerequisites, structures, characteristics, and modes of operation of matrix assemblies in the following. For more detailed research on matrix assemblies, we refer to the numerous literature sources in this chapter.

7.2.1 The Time Is Ripe: Enablers of Matrix Assembly

The idea of branched island production already existed before the first beginnings of an assembly line structure under Henry Ford. However, it was not until the last decade, in the context of many innovations in the field of Industry 4.0, that the technical prerequisites were created to implement matrix assembly.

While the increasing mechanization by means of steam power from the second half of the eighteenth century onwards was referred to as the first industrial revolution, the second revolution of industrialization was characterized by mass production and the use of electrical energy. The third phase saw an explosion of automation through innovations in electronics and IT. Personal computers, programmable logic controllers (PLCs), and software were now the basis of all successful productions. The fourth version number of the Industrial Revolution, Industry 4.0, describes the application of a variety of new technologies in electronics and software such as: artificial intelligence (AI), Internet-of-Things (IOT), big data analytics, cloud computing (Osterrieder et al., 2020; Culot et al. 2020). Ultimately, Industry 4.0 technologies will be used to manage the increased complexity of products and their production, supply chain uncertainties, and demand.

In the practical implementation of matrix assemblies, people often talk about cyber-physical systems (CPS), which are considered a prerequisite for implementation. In CPS, physical assets (equipment, sensors, etc.) are connected to digital processes in real time (Lee et al., 2015; Kusiak, 2018). Combining the real and digital worlds, decisions can thus be made in real time. CPS monitor current states, create process transparency, help make decisions and enable self-optimization of the system, through interaction among elements (Monostori et al., 2016).

In summary, the following innovations from the subject area of I4.0 could be seen as enablers of matrix assembly:

CPS—as Control System In matrix assembly, decisions have to be made in real time. CPS are the prerequisite to create a digital image, i.e., a digital twin, of the production, to connect it with other systems (logistics, maintenance, etc.) to make scheduling decisions, such as: "To which station will the order be transported next?" (Küpper et al., 2018).

Big Data Analytics and Artificial Intelligence Matrix production must be controlled in real time. With a multitude of unpredictable disruption possibilities such as material bottlenecks, machine failures, assembly disruptions, employee availability, and much more, it is impossible to pre-plan production for even one shift. On the contrary, one of the main propagated advantages of matrix assembly is the lossless handling of disturbances that would bring a classic assembly line to a standstill. Managing such a highly complex system in real time exceeds the capabilities of classic control algorithms. An independent continuous optimization based on past result data from the CPS will be another prerequisite for the successful operation of a matrix assembly (Hofmann et al., 2020).

Autonomous Transport Systems The further development of sensor technology, control systems, battery and charging technology, and the continuous cost reduction of automated guided vehicles enable products, materials, and resources to move independently in the matrix.

Robotics Innovations in robotics and handling systems enable matrix workstations to autonomously adjust to a variety of different products (Küpper et al., 2018).

Assistance Systems In a matrix assembly, a large number of different products and option variants are predominantly manufactured manually. This places increased demands on the information and qualification of the workers. A large amount of information must be provided digitally for specific orders and workers, depending on their qualifications. Screens, augmented reality (AR) and other projection technologies must be increasingly used (Küpper et al., 2018).

Research Results For almost 10 years now, numerous universities and research institutes, mainly in German-speaking countries, have been involved in researching and implementing matrix assembly. Very often in cooperation with local automotive companies. For example, the KIT in Karlsruhe, the TUM in Munich, the RWTH Aachen, the WHU in Vallendar, or the University of Braunschweig have contributed to analyzing the various challenges of matrix assembly and to offer practical solutions for the mostly NP-hard, i.e., complex and computationally intensive, optimizations.

7.2.2 Objectives of Matrix Assembly

The objectives of matrix assembly can be derived from the challenges in our introduction (Sect. 1.2). The following objectives are based on the works of Greschke et al. (2014), Kern et al. (2015), Greschke (2016), Bochmann (2018) and Hofmann et al. (2018):

- Production of a high variance of customized products and option variants
- Increasing utilization, and thus efficiency, in the manufacture of multi-variant products
- Integration of new products without major changes to the entire assembly and without a loss of production
- Lossless implementation of fluctuations in demand in volume and product mix without reconfiguration of the assembly system

If these four points are summarized and reduced to their core, the essential objective of matrix assembly is to reduce costs in the manufacture of multi-variant products by increasing flexibility and efficiency in assembly. For this purpose, the enablers from Sect. 7.2.1 are used, which, on the one hand, initially lead to an increase in fixed costs, but, on the other hand, make it possible to manufacture a wide variety of products in one assembly system.

7.2.3 Characteristics of Matrix Assembly

In contrast to the linear arrangement of the workstations in line assembly, the workstations in matrix assembly are mostly arranged in a grid—a matrix (Fig. 7.6). There can be different basic structures for this grid (Bochmann, 2018), e.g., these can be designed according to a fixed pattern. Free matrix structures, without a fixed grid structure and flow direction are also possible (Greschke, 2016). The layout can, but does not have to, have fixed defined entry and exit points (Hottenrott & Grunow, 2019). Based on this basic structure, Hottenrott and Grunow (2019) defined the Flexible Layout Design Problem (FLDP).

In order to use a matrix assembly, all work operations (WOs, see Sect. 2.6.2) must first be divided or grouped into work packages. This additional level is necessary because the work packages can be assigned to different stations in the matrix—an independent assignment of all WOs would be too complex, not very meaningful, and exacerbates the precedence relationships (Bochmann, 2018). This additional level in the planning process of the matrix assembly simplifies the later optimization problems in balancing or real-time control within the matrix. The work packages represent small, self-contained assembly contents. A detailed balancing of these work packages is not necessary or possible in the design phase of the matrix. Important parameters, such as the number of matrix elements, buffers in front of the matrix elements, and the distribution of the work packages among the cells, must be defined by means of analyses and simulations. In most concepts, the actual balancing takes place in real time through the launching sequence and the routing of the products by means of WOs in the matrix. These issues, i.e., the routing and the dimensioning of the matrix, represent the focus of current research (Hofmann et al., 2018). Efficient and effective routing in real time, including personnel planning, is currently one of the greatest optimization tasks and prerequisites for the successful use of matrix assembly (Hofmann et al., 2020).

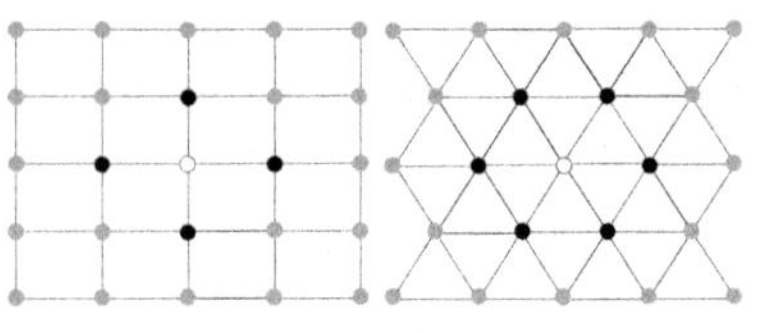

Fig. 7.6 Possible fixed basic forms of matrix assembly (adapted from Bochmann, 2018)

Material supply can be performed in a matrix assembly in different ways (Bányai et al., 2019; Filz et al., 2019). On the one hand, it is possible to design the material provisioning similar to the classic line assembly. Each matrix element contains the material necessary for the work packages assigned to it. However, space constraints in the individual matrix elements and resulting high material inventories limit this method of material provisioning. Manually or with the help of automated solutions, e.g., by using AGVs, the material could be brought to the matrix element just-in-time. Currently, there are approaches that automated material racks are flexibly transported into individual matrix elements. The worker removes the required material and the automated rack leaves the matrix element again. A third approach would be the use of kitting carts, which is also becoming increasingly common in line assembly. A cart with material kitted exactly for this order follows the product, either directly with the AGV of the product or on its own AGV.

It is obvious that in matrix assembly one cannot speak of a takt in the sense of a clocked flow assembly. Nevertheless, capacity planning of the matrix is required on the basis of an average takt. Very in-depth elaborations on this can be found in Greschke and Herrmann (2014) and Greschke (2016).

These characteristics can be summarized in the following Table 7.2 (Lenzen, 2020).

7.2.4 And Who Is the Winner: An Assembly Line in VarioTakt or Matrix Assembly?

A direct comparison with the wish of one clear winner is, of course, exaggerated. As in many areas of everyday business, there is no *one* better system. In the following comparison, however, we will be able to show that by using the VarioTakt, some of the disadvantages of classic mixed-model assembly in a fixed takt can be eliminated. Despite the high variance in the assembly line, we can shift the break-even point of matrix assembly back somewhat in the direction of an assembly line (compare Fig. 7.8).

7.2.4.1 Flexibility

To be able to evaluate flexibility, we use the structure of Sethi and Sethi (1990). The target variable which is important for the company and which needs to be increased is flexibility with respect to the market and thus flexibility in production. This is achieved through inherent system flexibilities, which in turn arise from three basic components. We modify the original presentation of Sethi and Sethi (1990) to better reflect our subject focus on assembly. Thus, we change the base capability “material handling” to “assembly systems” where we can distinguish mixed-model assembly in VarioTakt and matrix assembly (Fig. 7.7).

Table 7.2 Characteristics of matrix assembly (adapted from Lenzen, 2020)

Layout	Grid	The workstations are arranged in a standard grid, where each station can have four adjacent stations (Bányai et al., 2019; Greschke & Herrmann, 2014; Hottenrott & Grunow, 2019).
	Other	Workstations are arranged in a specific layout, e.g., triangular grid (Bochmann, 2018), circular, or sequential (Greschke, 2016).
Utilization	Full	Workstations are balanced to full or near-full utilization (Hottenrott & Grunow, 2019).
	Buffer time	The workstations are designed with a buffer time so that they are flexible in case of interruptions (Minguillon & Lanza, 2019).
Redundancies	Task boundary	Maximum number of task duplicates that may be entered into the system (Hottenrott & Grunow, 2019).
	Station boundary	Maximum number of tasks assigned to a single workstation (Greschke, 2016, p. 119).
Routing	Spontaneous	Routing decisions are only made when an order leaves the matrix cell, which implies autonomous optimization of routes (cf. Greschke, 2016). Ability to reschedule planned routes due to interruptions (cf. Bányai et al., 2019; Minguillon & Lanza, 2019).
	Planned	Scheduling and routing are based on planning. Rescheduling in case of disruptions or rush orders may or may not be allowed (Hofmann et al., 2018; Bányai et al., 2019).
	Integrated	The routing is taken into account in the balancing and is therefore predefined (cf. Hottenrott & Grunow, 2019) and will not be changed.
Flow direction	Free	Assembly object movement is unrestricted (Bányai et al., 2019).
	Targeted	The movement of the assembly object must follow a flow direction, vertical movements are possible, no backward flow is allowed (Hottenrott & Grunow, 2019).
Material handling	Cell-bound	The material is positioned in the matrix so that all assigned work packages could be processed.
	Order-bound	Material supply is organized in combination with order movement (with the product), e.g., via kitting carts (Filz et al., 2019).
	Independent	Materials are supplied via AGVs or similar systems that find their way through the system independently of the orders (Bányai et al., 2019).

In order to better compare the two assembly systems, we assume the basic flexibilities "machines/equipment" and "design/precedence relationships" to be constant.

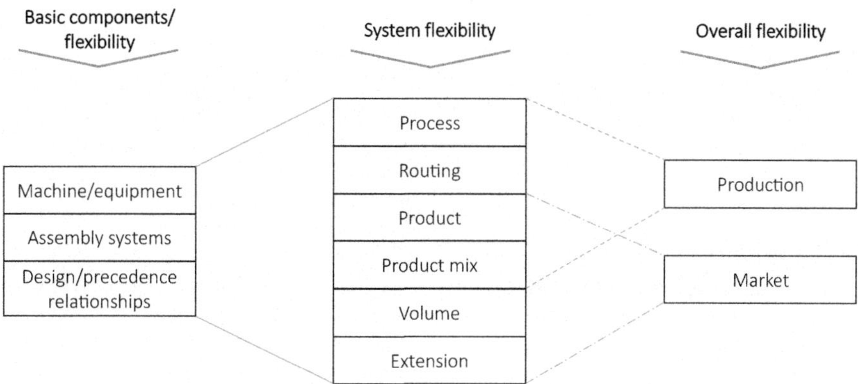

Fig. 7.7 Flexibility relationships for assemblies adapted from Sethi and Sethi (1990)

Matrix assembly	Mixed-model assembly
Process flexibility	
Is the core capability of matrix assembly; reactive rescheduling of routes and sequences are possible. Utilization can be maintained at a high level despite disruptions or changes in the model mix; reconfiguration of the system is not necessary.	Process sequences are defined in advance, there is no possibility to reactively adjust processes in case of disruptions. Reactively, cost-intensive additional workers (flexible workers) can be deployed in the event of disruptions or overload situations. → *VarioTakt*: no extended benefits
Routing flexibility	
Only the precedence relationships of the activities to be performed limit this flexibility. An identical product can use a wide variety of routes through the matrix.	No flexibility, a fixed route. In the event of disruptions, units can be discharged at defined points; in the vast majority of cases, this is associated with many disadvantages and is possible to a very limited extent. → *VarioTakt*: no extended benefits.
Product flexibility	
The assembly system can be adapted for new products or activities during ongoing production. Since identical activities can be performed on different assembly cells, individual cells can be deactivated and reconfigured.	The integration of a new product usually requires a complete redesign and synchronization of the assembly line. Even isolated additional activities can result in a modified balancing. → *VarioTakt*: Each product has its own independent line balancing and thus takt time. Changes to one product group have no effect on the other products in the line, and modification efforts are reduced.
Product mix flexibility	
The increased process and routing flexibility results in a significant increase in product mix flexibility. However, since individual products can take up specific resources, which are located in a limited number of assembly cells,	Since most mixed-model assemblies use the WATT balancing method (Sect. 2.6), series-sequence restrictions arise. When mix changes occur in the production program, an overload or underload may result in the need for modified line balancing.

(continued)

Matrix assembly	Mixed-model assembly
matrix assembly also has volume-mix restrictions.	→ *VarioTakt*: In the variable takt, the assembly line adapts to its load independently; despite changes in the model-mix, no re-balancing is necessary. Series-sequence restrictions and thus the sequencing problem (see Sect. 3.2.3) are significantly reduced.
Volume flexibility	
Can be easily achieved at short notice by changing work schedule or shift models. In addition, individual assembly cells can be flexibly deactivated or activated and thus volume capacity adjusted without changing the configuration of the system. The commitment to a fixed common shift start time for all workers can also be dissolved in this way.	Can easily be achieved in the short term by changing work schedule or shift models. For changes beyond this volume, the assembly line must be re-balanced. → *VarioTakt*: Volume changes resulting from mix changes are compensated. Beyond that, no further volume flexibility.
Expansion flexibility	
Can be successively expanded by installing new matrix elements, even during ongoing production.	An increase in the volume per unit of time can be achieved by a change of takt time and the associated increase in worker density. If an increase in worker density is no longer possible, the assembly line must be extended. This conversion of the entire line, which is usually very cost-intensive, is only possible during a production shutdown. → *VarioTakt*: no extended benefits.

Conclusion: Flexibility

In the area of flexibility, matrix assembly plays out its advantages, because this advantage is inherent in it. In terms of changing volumes, disruptions in the production process, or successive capacity expansion during ongoing production, it is clearly superior to the classic assembly line. By using the variable takt and the VarioTakt, the flexibility disadvantage of the assembly line can be significantly reduced in the event of product and product mix changes. In the end, a matrix assembly does not get by without restrictions in the mix of the production program; here, the VarioTakt is even superior due to its independence from the production mix. The type 3 mixed-model assembly in the VarioTakt is particularly impressive due to its low control effort compared to matrix assembly. The real-time control of (1) material, (2) product, and (3) personnel in matrix assembly usually ends up in three NP-hard optimizations which can only be solved in real time approximately and in isolation with complex heuristic procedures. Experienced optimizers often see this as the knockout criterion of matrix assembly, since heuristic methods are used for real-time solutions and there is a risk of getting stuck in local minima for one or all optimization/control problems.

7.2.4.2 Investments

In summary, it can be said—and all previous publications currently agree on this—that the investment costs of matrix assembly are much higher than those of assembly lines (Küpper et al., 2018; Hottenrott & Grunow, 2019; Kampker et al., 2021).

Production Area On the one hand, matrix assemblies have an increased space requirement per station and need significantly more transport areas between the cells, as we explain in Sect. 7.1.3. On the other hand, simulations showed that despite a lower number of workers, the total number of workstations is higher than that of a comparable assembly line (Greschke et al., 2014; Greschke, 2016; Bochmann, 2018).

Transport Technologies Matrix assemblies require free routing by means of AGVs; the costs of these systems are still significantly higher than the costs of a classic chain conveyor or apron conveyor today—even if this difference is expected to shrink significantly in the next few years. The variable takt does not presuppose the use of an AGV system; there are much lower-investment alternatives, as we have described in Sect. 4.6.

CPS and Control Systems Operating and managing a matrix assembly requires deep embedding of CPS and elaborate automated control logic. In many cases, integration of artificial intelligence (AI) technologies or self-learning systems is already assumed. The balancing of a matrix assembly, if one can speak of balancing at all in this case, initially requires less effort than in the case of a line assembly. In return, the control effort in the operation of a line assembly is significantly reduced or, with the exception of series-sequencing, does not exist. The sensor technology and IT infrastructure of a CPS are further investment drivers, but the comparison with line assembly is invalid in this case, since these assembly systems will also increasingly use CPS for optimization in the future.

Equipment In order to fully exploit the benefits of matrix assembly, it is necessary to be able to perform identical activities in different matrix elements. Thus, in contrast to an assembly line, these resources must be purchased several times.

Conclusion: Investments

In principle, it can be assumed that the initial investments of a matrix assembly are significantly higher than those of a line assembly, as the previous examples have shown. If an assembly line is operated with the VarioTakt, investments can be avoided despite increased flexibility with reduced running costs due to lower takt losses.

7.2.4.3 Operating Costs

Takt Losses One of the main arguments for matrix assembly and criticism of line assembly is the handling of large takt time spreads resulting from a high variance of and in the units to be produced. Most of the current publications focus on analyzing this advantage of matrix assembly (Kern et al., 2015; Greschke, 2016), for example, in a study by the Boston Consulting Group (Küpper et al., 2018), worker utilization could be increased by 11%. Kampker et al. (2021) show a 19% increase in efficiency when integrating hybrid assembly in the assembly of automobiles with hybrid and conventional drive systems. However, these advantages are not compelling; Hofmann et al. (2018) initially see lower utilization in the matrix compared to the line. In Chap. 4, and specifically in Sect. 4.5, we showed the benefits of using variable takt. Takt time spreads, caused by workload differences between different products (not option variance, Sect. 2.4), are almost completely absorbed by the variable takt—there are *no takt losses due to different products and their takt time spread in the assembly line!* From the point of view of takt losses in an assembly of different products, variable takt and the VarioTakt are at least equal to matrix assembly, with lower expenditure in operational control. In the case of selective takt time spreads, triggered by special option variants, the matrix can show its advantages and avoid takt and model-mix losses.

Utilization Losses Reacting in real time is one of the other main arguments in favor of matrix assembly. While the complete line assembly grinds to a halt when a fault occurs at just one station, the disruption or failure of one matrix element is compensated by other elements. A proof of this advantage by simulation is provided by Hofmann et al. (2018).

Furthermore, the performance level of each cell can be adapted to the respective worker; if a matrix cell is staffed by a worker with reduced performance, this is taken into account in real time in the control system without significantly reducing the output of the entire system. This is not possible in line assembly. Using performance-reduced workers in a flowing assembly line is in itself a challenge. The station in question must be provided with a lower performance level in the line balancing, and in operation, the utilization of the station is then always lower, regardless of whether it is staffed by a performance-reduced worker or not.

However, Hofmann et al. (2018) also recognize in their work that flow assembly lines can already have a very high availability. An availability of over 95% is no longer a rarity for a professionally managed and continuously improved assembly line. Already in Sect. 2.1 we showed the managerial advantages of flow production, which brings problems to the surface and forces perfection—"failures must cause pain", otherwise there is no need to eliminate the causes lastingly. So, it is not surprising that we consider the benefits of matrix assembly for avoiding losses in case of failures to be rather low. Mind you, this hypothesis comes from our practical experience, scientific empirical studies are still missing. A variable takt assembly line thus takes advantage of all the managerial benefits of flow production. We estimate the benefit of the matrix by avoiding losses during disruptions as low … perhaps it is even counterproductive, since it does not force the causes of the problems to be eliminated in the long term.

Depreciation Due to the increased investments, the depreciation costs of a matrix assembly are also higher compared to a classic line assembly. Since, in addition to the goal of higher flexibility, another focus is the increased utilization of all workers, it is necessary to maintain a reserve of matrix cells, including equipment and resources, in order to be able to staff them with workers at any time. The order and the matrix cell should wait for the worker, not the worker for the order. This means that the utilization of the matrix, and thus the investment utilization, is lower than in line assembly.

CPS and Control Systems Since the control efforts of a matrix assembly are significantly higher than those of an assembly line and end up in NP-hard optimizations, it can be assumed that the ongoing operation, maintenance and regular adaptation of these systems are significantly more complex and thus more cost-intensive.

Conclusion: Operating Costs
Defining a clear winner between the two systems in terms of operational costs is the least clear of all the comparisons. What is clear is that the more products that need to be manufactured with different total assembly workloads and large takt time spreads, the use of matrix assembly becomes more advantageous in terms of operational costs. By using variable takt in an assembly line, one of the biggest advantages of matrix assembly, the minimization of takt losses, is significantly reduced, the break even point is pushed quite a bit towards an assembly line in VarioTakt (cf. Fig. 7.8).

7.2.4.4 Complexity and Controllability

The significantly higher degrees of freedom in the assembly sequence, the compensation of disturbances in real time, and a control system trimmed to ideal utilization of the workers require the use of highly complex control software, combined with elements of artificial intelligence and learning algorithms. Such a system is actually impossible for the responsible management to steer independently; cognitive improvements to the assembly system can no longer be carried out without elaborate simulations. If a failure or even a malfunction of the control system occurs, the

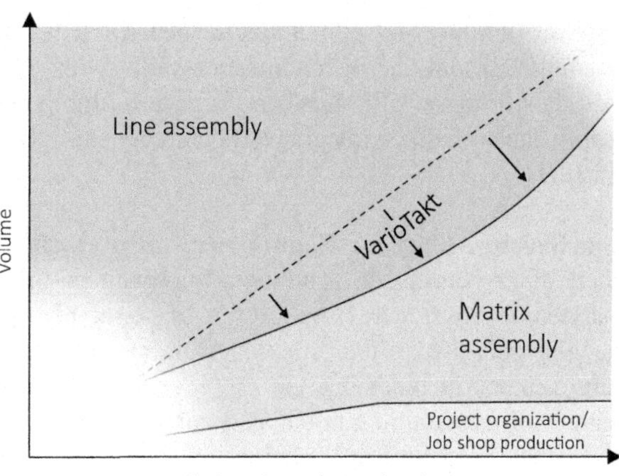

Fig. 7.8 VarioTakt—an extension of the range of use of line assembly

matrix assembly system is incapable of action. Adjustments or the elimination of error sources in the control of an assembly line can still be observed, analyzed, and changed by planners and management—this is no longer possible in matrix assembly.

Conclusion: Complexity and Controllability

An assembly line impresses with its simplicity of control in contrast to matrix assembly. The variable takt increases the complexity of the assembly line somewhat, but this is more than compensated for by the advantages. But let us' not kid ourselves, the time when an individual could analyze each (assembly) control system iteratively is over. New, learning systems have their justification in future production control, for matrix assembly this is necessary, management will have to adapt to it.

7.2.4.5 Ability for Continuous Improvement

The advantage of an assembly line is obvious here, often simple observations, in the Toyota Production System (TPS) "go to gemba," is the best method to generate improvement ideas. All interrelationships are linear and the same in every run. These are the best conditions for continuous improvement with the participation of the employees concerned. In an assembly line, the cause and effect of malfunctions or quality defects coincide very closely in time and always run in exactly the same sequence; in the matrix, this need not be the case. The assembly sequence of two identical orders may differ in the assembly matrix. A root cause analysis in case of quality problems could prove to be very difficult due to different assembly sequences. Different assembly sequences could make root cause analysis for defects of any kind much more difficult.

In the complex system of matrix assembly, on the one hand, inefficiencies are much more difficult to detect by the individuals involved. New forms of cooperation in the optimization of production between workers, planners, and managers must be found and require higher qualifications at all levels.

On the other hand, if learning algorithms deliver what they promise, the matrix system will improve itself without the need for a motivated individual. Since matrix assembly can only be operated successfully with significantly more collected and used data from the CPS, this can generate further potential. For example, the entire supply chain could be integrated much more strongly and deeply into the assembly system.

Conclusion: Ability for Continuous Improvement

Many of the concepts of continuous improvement that have been perfected over the past decades must be revised for a matrix assembly. In contrast to matrix assembly, the assembly line offers simple conditions for the analysis and application of improvement methods by the employees involved. Variable takt increases the degrees of freedom in continuous improvements due to the independent balancing of products. Since the interrelationships in a matrix assembly can literally no longer be seen through by a single individual, simulations will have to be increasingly used.

In our opinion, this is not a disadvantage, but an evolutionary development in the management of assembly systems.

7.2.5 Outlook Matrix Assembly

The majority of publications on matrix assembly come to the common conclusion that matrix assembly can show its advantages at low volumes combined with high variance in products and options (Kern et al., 2015; Greschke, 2016; Hofmann et al., 2018, Kampker et al., 2021). The manufacturing industry currently still owes a real POC (proof of concept)—pioneers in a comprehensive application of matrix assembly are being sought!

If the market is characterized by continuously increasing or extremely fluctuating demand, matrix assembly could be superior to classic assembly line due to its continuously expandable or reducible capacity (activate and deactivate matrix elements). If the market development cannot be estimated, it could be started with a small matrix at first; if the demand develops positively, the matrix can be continuously expanded while production continues (Küpper et al., 2018). An assembly line that is designed too large from the start causes unjustifiable costs. If it is designed too small, an expansion is often associated with high costs, investments, and a production interruption.

In the production world with increasing variance and product individualization, matrix assembly will find its justification or its permanent place. Improved methods for mastering and continuously improving matrix assembly will have to evolve. Priority should be given to real-time methods that avoid ending up in local optima. Management concepts suitable for a matrix, e.g. for the CIP process, should be developed, and the costs of the infrastructure must be further reduced. An expansion of the application of matrix assembly can be expected. Nevertheless, the assembly line should not be written off too quickly (Fig. 7.8).

We were able to show that variable takt can implement many of the advantages of matrix assembly also in an assembly line, and this with less effort in implementation, control and while retaining the advantages of flow production. Figure 7.8 ranks both systems with the variables volume and variance and illustrates the impact of VarioTakt in line assembly. Future research could attempt to quantify these limitations, but a real breakthrough in the use of matrix assembly will require pioneers in the manufacturing industry. Some first applications or adaptations are briefly presented in the next chapter.

References

Bányai, Á., Illés, B., Glistau, E., Machando, N., Tamás, P., Manzoor, F., & Bányai, T. (2019). Smart cyber-physical manufacturing: Extended and real-time optimization of logistics resources in matrix production. *Applied Sciences, 9*(7), 33 p. https://doi.org/10.3390/app9071287

Bochmann, L. S. (2018). *Entwicklung und Bewertung eines flexiblen und dezentral gesteuerten Fertigungssystems für variantenreiche Produkte* (185 p.). ETH Zurich. https://doi.org/10.3929/ethz-b-000238547

Culot, G., Nassimbeni, G., Orzes, G., & Sator, M. (2020). Behind the definition of industry 4.0: Analysis an open questions. *International Journal of Production Economics, 226*. https://doi.org/10.1016/j.ijpe.2020.107617

Filz, M.-A., Herrmann, C., & Thiede, S. (2019). Analyzing different material supply strategies in matrix-structured manufacturing systems. *Procedia CIRP, 81*, 1004–1009. https://doi.org/10.1016/j.procir.2019.03.242

Greschke, P. (2016). *Matrix-Produktion als Konzept einer taktunabhängigen Fließfertigung* (180 p). Vulkan Verlag.

Greschke, P., & Herrmann, C. (2014). Das Humankapital einer taktunabhängigen Montage. *ZWF Journal of Economic Factory Sales, 109*(10), 687–690.

Greschke, P., Schönemann, M., Thiede, S., & Herrmann, C. (2014). Matrix structures for high volumes and flexibility in production systems. *Procedia CIRP, 17*, 160–165. https://doi.org/10.1016/j.procir.2014.02.040

Hofmann, C., Brakemeier, N., Krahe, C., Stricker, N., & Lanza, G. (2018). The impact of routing and operational flexibility on the performance of matrix production compared to a production line. In R. Schmitt & G. Schuh (Eds.), *Advances in production research* (pp. 155–165). https://doi.org/10.1007/978-3-030-03451-1_16

Hofmann, C., Krahe, C., Stricker, N., & Lanza, G. (2020). Autonomous production control for matrix production based on deep Q-learning. *Procedia CIRP, 88*, 25–30. https://doi.org/10.1016/j.procir.2020.05.005

Hottenrott, A., & Grunow, M. (2019). Flexible layouts for the mixed-model assembly of heterogeneous vehicles. *OR Spectrum, 41*, 943–979. https://doi.org/10.1007/s00291-019-00556-x

Kampker, A., Kawollek, S., Marquard, F., & Krummhaar, M. (2021). Potential of hybrid assembly structures in automotive industry. *Aachen University* Library, 10 p. https://ssrn.com/abstract=3848756

Kern, W., Rusitschka, F., Kopytynski, W., Keckl, S., & Bauernhansl, T. (2015). Alternatives to assembly line production in the automotive industry. *23rd International Conference on Production Research* (9 p.). http://publica.fraunhofer.de/eprints/urn_nbn_de_0011-n-3798198.pdf

Küpper, D., Sieben, C., Kuhlmann, K., & Ahmad, J. (2018). Will flexible-cell manufacturing revolutionize Carmaking? *The Boston Consulting Group Online*, October 18. Accessed April 28, 2021, from https://www.bcg.com/de-de/publications/2018/flexible-cell-manufacturing-revolutionize-carmaking

Kusiak, A. (2018). Smart manufacturing. *International Journal of Production Research, 56*(1–2), 508–517. https://doi.org/10.1080/00207543.2017.1351644

Lee, J., Bagheri, B., & Kao, H. A. (2015). A cyber-physical systems architecture for industry 4.0-based manufacturing systems. *Manufacturing Letters, 3*, 18–23. https://doi.org/10.1016/j.mfglet.2014.12.001

Lenzen, D. A. (2020). Industry 4.0 – Potential and challenges of matrix production. *WHU Otto Beisheim School of Management.*

Minguillon, F. E., & Lanza, G. (2019). Coupling of centralized and decentralized scheduling for robust production in agile production systems. *Procedia CIRP*. https://doi.org/10.1016/j.procir.2019.02.099

Monostori, L., Kádár, B., Bauernhansl, T., Kondoh, S., Kumara, S., Reinhart, G., & Ueda, K. (2016). Cyber-physical systems in manufacturing. *CIRP Annals, 65*(2), 621–641. https://doi.org/10.1016/j.cirp.2016.06.005

Osterrieder, P., Budde, L., & Friedli, T. (2020). The smart factory as key construct of industry 4.0: A systematic literature review. *International Journal of Production Economics*, 221 p. https://doi.org/10.1016/j.ijpe.2019.08.011

Sethi, A. K., & Sethi, S. P. (1990). Flexibility in manufacturing: A survey. *International Journal of Flexible Manufacturing Systems, 2*, 289–328. https://doi.org/10.1007/BF00186471

Advanced Concepts in Automotive Manufacturing to Master Variance in Assemblies

8

In this chapter, we would like to present further concepts for mastering variance in assemblies. In addition to a comparison with VarioTakt, we can show that all concepts can be combined with variable takt and thus their range of action can be extended even further. In the main focus, we will use variable takt in AD for MMA and thus in the phase of balancing the assembly line. At the end of this section, we will outline an additional idea of how variable takt could also be used in SD for MMA, i.e., in workload-oriented series-sequencing.

8.1 Modular Assembly at Audi

One of the pioneers of matrix assembly, both in cooperation with universities for basic research and in communicative dissemination in industry, was and is Audi. For more than 6 years, the company has been working on the industrialization of Modular Assembly, as matrix assembly is called at Audi. Spin-offs, such as the Arculus GmbH, have already resulted. Arculus, a company initially focused on control software for matrix assembly, has now also entered the market for mobile robots, because integrating the software into its own robots and automated transport systems is much easier and more targeted to implement. Audi and Arculus are not concerned with automating the actual assembly activity. The target is to automate all supporting transport activities of products and assembly material as well as the predictive and reactive control of the entire matrix in real time.

8.1.1 Objectives and Challenges of Modular Assembly at Audi

With Modular Assembly, Audi essentially wants to utilize the advantages that we described in Sect. 7.1.2. Specifically, these are (Kern et al., 2015):

P. Bebersdorf, A. Huchzermeier, *Variable Takt Principle*, Management for Professionals, https://doi.org/10.1007/978-3-030-87170-3_8

1. As many advantages as possible of flow assembly (cf. Sect. 2.1) should be retained by maintaining a division of labor and allowing workers to concentrate exclusively on value-adding processes.
2. Logistics and assembly are to be viewed as an overall system. Despite a strict separation of value-adding and non-value-adding processes, the entire value stream is to be optimized in Modular Assembly.
3. The overall system should be oriented toward the employee and thus lead to significantly more ergonomic and attractive workplaces (Greschke et al., 2014). The workload is to be distributed in a much more harmonized manner and adapted to the individual skills and performance capabilities of the workers.
4. Flexibility in operation and in the event of modifications to the assembly system is to be significantly increased.
5. Losses in the event of disruptions or rescheduling of the system are to be minimized. The assembly system shall be significantly more resistant to volume and product-mix variations. Series-sequence or mix restrictions shall be eliminated.

Challenges

These advantages are offset by significant challenges. On the one hand, this would be a very elaborate and complex control system for Modular Assembly, which we have already discussed in Sect. 7.2.1. This not only involves the control of the assembly objects and the material but also the staffing and the distribution of the workers must be controlled in real time. The spin-off of the already mentioned Arculus was a reaction to this.

The integration of logistics and the associated space requirements, which represent one of the largest investments, are difficult to integrate into existing (building) structures.

The entry and exit times into the individual matrix elements or cells represent a loss. In the individual cells, stationary assembly takes place. When the assembly object enters and leaves the cell, the worker cannot add any value. A theoretical solution would be to design the cells, as a flow system. The assembly object moves continuously through the matrix element, when the worker moves to the next assembly object there would always be two products in the matrix element, the worker can move to the next product without losses. In order to maintain the ability of an individual station time adapted to the assembly object, this continuous outfeed and infeed at constant speed could be interrupted by an order-individual stop in the middle of the station.

Perhaps the greatest challenge in implementation is shared by Modular Assembly and variable takt (cf. Sects. 4.8 and 4.9)—it is a matter of convincing management of the merits of a new assembly system that has not yet been proven in practice. Pioneers are necessary—Fendt is one of the pioneers in variable takt, Audi is one of the pioneers in Modular Assembly.

8.1.2 Nine Principles of Modular Assembly Concept in Comparison with VarioTakt

In their work, Kern et al. (2015) describe nine essential principles of Modular Assembly, which we would like to compare in the following to an assembly line with variable takt. The principles and the comparison are similar to the explanations in Sect. 7.2.4.1 "In comparison: flexibility" and build on them, but are not identical.

Modular Assembly (Kern et al., 2015)	Line assembly in VarioTakt
Principle 1: Variable assembly sequence	
Only the precedence graph of the assembly activities specifies restrictions in the sequence of the assembly activities. Components of identical orders can thus be assembled in different sequences. → The variable assembly sequence is not an advantage in itself, but the prerequisite for principles 2, 4, and 5.	No flexibility, a fixed assembly sequence is specified. → Fixed assembly sequence can be a distinct advantage in root cause analysis of quality defects. →As a result, no real-time control effort is required during operation of the assembly system.
Principle 2: Variable assignment of the subsequent station	
Due to the variable assembly sequence, each order can pass through the cells in an individual sequence; cells can be skipped. Taking into account the restrictions of the precedence graph, the cell with the shortest waiting time is selected as the next cell. → Is a prerequisite for principle 4.	The next station is fixed in the assembly line and cannot be changed. → As a result, no real-time control effort is required in the operation of the assembly system.
Principle 3: Variant-dependent station or processing time	
The assembly objects remain in the station only depending on their individual assembly time. → Maximum or minimum variants in product or options no longer lead to losses in utilization (cf. Sect. 2.8). → Optimizations of individual assembly times lead directly to efficiency gains without the need for modified line balancing.	In the VarioTakt, each product receives its own takt time; option variance is controlled with the WATT up to a defined limit β. → Variance between products does not lead to any loss of utilization (cf. Sect. 2.8). → Utilization losses, due to variance in options, are reduced.
Principle 4: Compensation of disturbances	
If individual cells are disrupted and temporarily unavailable, this could be the case in the event of a technical failure or a quality deviation, the entire assembly system does not stop. This is made possible by redundant cells and the variability in the precedence graph. → Losses due to local disturbances are not transferred to the entire system.	The malfunction or stop of a station leads to the stop of the entire system, possibly reduced by buffers. This comparison with matrix assembly is perhaps one of the most controversial, because: → From the point of view of Lean Management, it can make more sense not to compensate for the effects of faults. In this way, the compulsion to permanently eliminate the cause of the faults is not suppressed; we describe these effects in Sects. 2.1.1 and 2.1.3.

(continued)

Modular Assembly (Kern et al., 2015)	Line assembly in VarioTakt
Principle 5: Compensation product-mix changes	
Individual cells can be activated or deactivated. The assignment of the matrix cells is continuously adapted to the model-mix currently present in the system. → Workload changes due to changes in the model-mix of the production program are compensated and do not lead to model-mix losses (see Sect. 2.8.2). → Effort, by forcing the system to be re-balanced, is avoided.	Mastering product variance is one of the core capabilities of the VarioTakt. → Model-mix losses are avoided. → No control effort is required in the operation of the assembly system.
Principle 6: Incremental changes during operation	
Modifications to the assembly system caused by new products, volume, or mix changes can be made incrementally during ongoing production. → Fast adaptability of the system. → No loss of production during modifications.	In VarioTakt, individual product groups can be balanced independently of each other (see Sect. 4.5). → Minimize the effort of line balancing when a product changes in the portfolio. → Re-balancing due to mix changes are not necessary. → Structural modifications to the assembly line require a production interruption.
Principle 7: Transports by AGV	
The assembly objects and the assembly material are autonomously brought to the worker by AGVs. → Prerequisite for principles 2 and 4.	A prerequisite for the use of variable takt is a conveyor system with adjustable distances between the assembly objects (cf. Sect. 4.6). → Requirements for the transport system slightly increased compared to the assembly line with fixed takt, but significantly reduced effort compared to the conveyor system of the matrix assembly.
Principle 8: Integrated quality controls and rework	
Quality checks and any necessary rework can be carried out at any point in the process without having to stop the entire system or laboriously disassemble the completed object. The station time can be reactively extended for this purpose or the assembly object can be transported to a rework area/cell if required. → Fast troubleshooting, only defect-free orders move to the next process step. → Reduced rework effort.	Rework can only take place at the end of the assembly line or the assembly line must be stopped. → Comparable discussion as to principle 4.

(continued)

Modular Assembly (Kern et al., 2015)	Line assembly in VarioTakt
Principle 9: Station times individually adapted to employees	
Station times for employees to be trained or employees with performance limitations can be customized. → Losses due to on-the-job training are not transferred to the entire system. → The heterogeneity of the workforce and demographic change can be taken into account, cost reductions through the integration of these employee groups.	The workload on the individual assembly stations is set in the line balancing. → Although a station could be specifically set with a lower utilization, this lower utilization is then always present, regardless of which worker is staffing the station.

One, in our view, significant advantage of matrix assembly is not addressed by Kern et al. (2015). In an assembly line, a fixed labor capacity of the same size is required at any time—or in other words: the number of workers required is the same at any time. In matrix assembly, it is possible to adapt the output of the system to the available worker capacity. Individual cells can be activated or deactivated, so that, for example, production can be maintained despite an increased level of sick leave. Or it is possible to produce at lower capacity during vacation periods by deactivating cells. A fixed shift start or end could also be dispensed with, and the capacity of Modular Assembly could be changed fluidly throughout the day.

8.1.3 Conclusion: Modular Assembly

Audi is one of the key pioneers in the introduction of matrix assembly and has thus clearly supported the spread of this assembly system. The first successful applications are running with the support of Arculus and other system suppliers in the assembly of electric motors at Audi's Hungarian plant in Győr. While Audi initially focused its activities on representing complete automotive assemblies using a matrix structure, hybrid systems have now also emerged in which line elements are combined with matrix elements. This is also the case with the assembly of electric drive units at the Győr plant. We were able to show that line assembly in VarioTakt can compensate for many advantages of Modular Assembly, the four most significant remaining advantages of Modular Assembly being:

1. The flexible integration of the heterogeneous workforce
2. The possibility of immediate rework
3. The control of large local takt time spreads of option variants
4. The reaction to disturbances

Whereby the latter point can be discussed controversially in terms of Lean Management. Many advantages in the assembly of products with different overall assembly workload and high option variance can also be implemented with the VarioTakt, with certainly much lower investment and operating costs and a significantly minimized complexity of the overall system. This comparison of Modular Assembly with the VarioTakt shows that there is no clearly better or "the right" assembly system. So, as always, the decision is up to the responsible management. It is not always necessary to decide in favor of one or the other system; a smart combination of both approaches is described in the following section.

8.2 Hybrid Assembly at BMW - Flex Segments

Matrix or line—which is the best system to assemble an automobile? Both—that is how the German car manufacturer BMW would answer this question. BMW is intensively involved in the research and practical feasibility of hybrid assembly structures. Specifically, with LML structures, as we presented them in Chap. 7. First approaches of hybrid assembly structures were described in the work of Hottenrott and Grunow (2019) and later elaborated in a practical way by Kampker et al. (2021). If current publications on hybrid assembly concepts always use multidimensional matrix elements in the same way, we would like to briefly introduce the simplest form of a hybrid assembly here. It is the LML or LM structure with a one-dimensional M1 matrix (Fig. 8.1). We have explained the operation of the LM structure in Sect. 7.1 using the example of Fendt. In the LML structure with an M1 matrix, the assembly line is interrupted at one point and a matrix element is inserted. If an order is moved into the matrix element, an order from the matrix element must be moved back into the line at the same moment in time so that no gaps occur in the assembly line. Not every order has to pass through the matrix element. The M1 matrix can consist of several parallel stations to increase the stationary takt in the matrix element. The more cells, the higher the stationary takt time that can be covered. In the LML structure, the series-sequence of the downstream line is changed, but the new sequence of orders after the matrix element can be

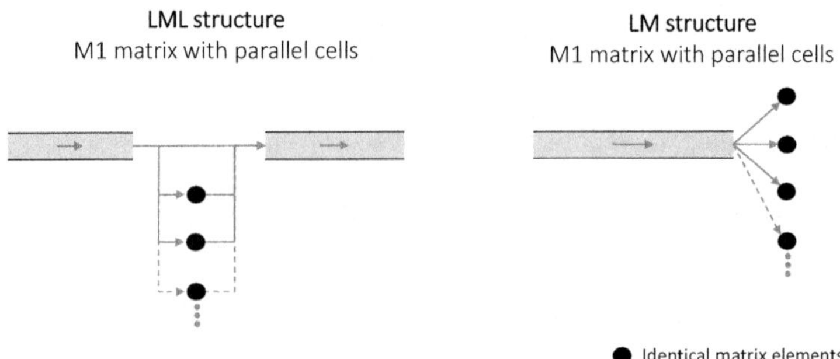

Fig. 8.1 The simplest hybrid assembly structure—LML and LM structure with a M1 matrix

predetermined. This modified but predetermined sequence of assembly objects can be used to plan and control all downstream logistics processes. Reactive real-time control in case of disturbances in the matrix or assembly line is not provided. A malfunction or stop in the matrix element causes the assembly line to stop.

8.2.1 Objectives and Mode of Operation of Hybrid Assembly with Flex Segments at BMW

In a study of assembly experts from the automotive sector, 58% of the respondents considered the hybrid assembly system to be the most promising way to master the increasing number of variants and to process them efficiently (Kampker et al., 2019). Only 27% believed that line assembly would be able to do this, and 15% that island assembly would. The same study asked for the best approaches to reduce takt losses within assembly planning, Fig. 8.2 shows the results. The outsourcing of variant-specific content (40%) and a more flexible workforce planning (27%) were attested to have the greatest leverage. In the following, we will deduce that exactly these two approaches can be implemented very well with hybrid assembly structures.

The research area of hybrid assembly structures is based on the following requirements for future assemblies (Kampker et al., 2021):

- Technical and economic ability to deal with a high number of variants
- Loss of utilization should be avoided, without increased production space or investment requirements

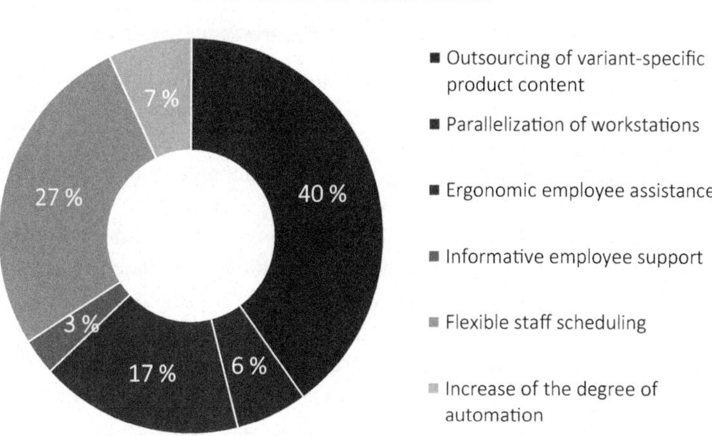

Fig. 8.2 Study results—assembly planning approaches to reduce workload losses (adapted from Kampker et al., 2019)

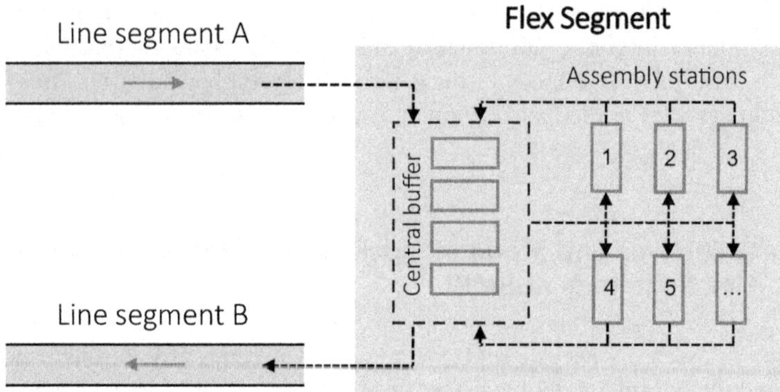

Fig. 8.3 Functional schematic diagram of hybrid assembly system - Flex Segments (adapted from Kampker et al., 2021)

- Increased utilization of line segments as variance is shifted to Flex Segments (Fig. 8.3)
- Increasing the degrees of freedom for product, product-mix and volume changes, as future developments cannot be accurately forecast
- Advantages of short learning curves (with low training expenses) through division of labor and standardization are to be retained
- Stable series-sequencing, for supply chain effectiveness and efficiency

The biggest difference of these requirements compared to matrix assembly is the intention to maintain a stable and predetermined sequence of assembly objects. A stable supply chain and high efficiency in the logistics processes are the motivation for this requirement. Later in this section, it will be seen that this requirement results in much simpler control of the entire system. A reaction, i.e., a rescheduling in real time in case of disturbances, is not desired, but also no longer possible.

The concept of hybrid assembly at BMW is based on the LML structure. Two line segments are interrupted by a multidimensional matrix, input and output buffers are combined to save space. Vehicles from the incoming assembly line are first temporarily stored in the buffer and then pass through none, one or more matrix elements with an order-specific stationary takt time. The combination of a central buffer and a matrix is called Flex Segment (Kampker et al., 2021). Except for the entry and exit times into the cells, the workers can thus be fully utilized. Neither takt nor model-mix losses occur in the cells. A possibility to avoid the loss of the entry and exit times has been described at the end of the previous Sect. 8.1.1. The individual cells can be activated or deactivated according to capacity requirements, which allows flexible worker control. The entire sequence is pre-planned, so that the input and output sequences are different but predetermined. The sequence in or between the cells is not predetermined and can follow different vehicle-station assignment control logics (Kampker et al., 2021). Disturbances in the cells are not compensated and can thus, mitigated by the buffer, be reflected as stops in the overall assembly line.

For the enablers, objectives, and characteristics of the cells of the Flex Segment, the same findings apply as for the matrix assembly and are described in Sects. 7.2.3.

Figure 8.3 depicts a LML structure with a multidimensional matrix. In the concepts for initial practical applications, BMW focuses on simple LML structures with few cells; these elements could be called "mini-matrix." The reason for this decision is the simplified control and lower complexity of these hybrid units.

8.2.2 Core Elements and Basic Principles of Flex Segments

BMW's hybrid assembly concept is based on two core elements (Kampker et al., 2021):

1. *Segmentation*: Assembly contents are already assigned to the Flex Segment(s) and the line segments in the planning phase of the assembly system. All variant-unspecific assembly contents remain in the assembly line. Option variants with large time spreads are assigned to the Flex Segment(s). In anticipation of a combination of hybrid assembly with variable takt, all product-specific content that differs significantly from the other products (local takt time spreads) should also be assigned to the Flex Segment(s). For example, the installation of the battery of an electric vehicle, in a mixed-model assembly with combustion vehicles, is predestined for assignment to the Flex Segment(s). If electric vehicles become established, this allocation could be reversed in the future and the combustion engine could be installed in the Flex Segment(s).
2. *Fixed line sequences/flexible matrix control*: The series-sequences of the individual line segments differ, but are predefined per line segment. The fixed sequences of the line segments are the prerequisite for supplying the assembly with JIT and JIS components, without which the material supply of an automotive assembly would be unmanageable. Despite this fixed input and output sequence, the control in the Flex Segment(s) is not predetermined. For this reason, the authors recommend the use of kitting carts within the Flex Segment(s).

Kampker et al. (2021) further describe nine basic design principles of a hybrid assembly:

• Variance concentration	In order to minimize the number of Flex Segments, variant contents should be able to be combined and processed in groups as far as possible.
• Decoupling	A central buffer per Flex Segment decouples lines and matrix elements from each other.
• Skipping	Vehicles only pass through the cells of the Flex Segments if the variant content requires it.
• Individual stationary takt	The station time in the individual matrix cells is order-specific individual but predetermined.
• Directed material flow	In both the line and Flex Segments, the material flow is directed.

(continued)

• Parallel processes	In order to minimize the effort in the control and to ensure the directed material flow, the cells of the Flex Segments are approached individually from the buffer. If an order requires the passage of two cells, it is again routed via the buffer.
• Independence product-mix	To minimize investment and reconfiguration costs, only content that is expected to experience high utilization losses in the long term will be moved to Flex Segments. Seasonal effects are not to be absorbed.
• Flexible product transport	Within the Flex Segment, the units are to be transported on an AGV, or they move independently into the cells. The assembly line can use a rigid transport system (e.g., apron conveyor, conveyor chain).
• Extensibility	When integrating new products, changes in the product-mix or volume, the Flex Segments can be expanded or reduced. This, compared to the assembly line, with significantly less reconfiguration effort, since changes in the Flex Segments can be installed and commissioned in parallel with operation.

In the same paper, Kampker et al. (2021) showed in a simulation with real data from the assembly of BMW vehicles with combustion engines and plug-in hybrids, an increase in line utilization of +19% with a simultaneous reduction of 4 line workstations when using a Flex Segment.

8.2.3 Hybrid Assembly with Flex Segments in Combination with VarioTakt

It is unnecessary to put the concepts of VarioTakt and hybrid assembly in competition with each other, because a combination of the two approaches seems much more interesting. This is because the strengths of one concept compensate for the weaknesses of the other. On the one hand, the VarioTakt can cope very well with workloads of different sizes but evenly distributed of the assembly line for the individual orders (global takt time spreads)—this is its strength, hybrid structures do not help here. Point concentrations of workload (local takt time spreads), on the other hand, can be handled very well with hybrid assemblies, this concentration is a prerequisite for using hybrid structures. If we combine both approaches and evaluate them with the same criteria from Sect. 7.2.4, we obtain the following findings.

8.2.3.1 Flexibility

To be able to evaluate flexibility, we again use the structure of Sethi and Sethi (1990), which we had already used in Sect. 7.2.4.1.

Advantages due to hybrid assembly:	+	Advantages due to mixed-model assembly line in VarioTakt:
Routing flexibility		
Possible in the Flex Segments, excluded in the line segments.	+	No flexibility, a fixed sequence.
Process flexibility		
Is excluded in the line segments, reactive additional flexible workers can be used in overload situations.		
Due to the free routing, the cells can be approached in different sequences.	+	
Product flexibility		
If the product modification can be concentrated in Flex Segments, no changes are necessary in the assembly line. Flex Segments can be adapted for new products or activities during ongoing production.	+	Each product has its own line balancing and thus its own variable takt; changes in one product thus have no effect on the other products in the assembly line; re-balancing efforts are reduced.
Product-mix flexibility		
If product-mix changes lead to an increase or decrease in workloads for the Flex Segments, then cells can be activated or deactivated.	+	In the variable takt, the assembly line adapts to its workload independently; despite changes in the model-mix, no re-balancing is necessary. Series-sequence restrictions and thus the sequencing problem (Sect. 3. 2.3) are significantly reduced.
Volume flexibility		
Can be achieved in the short term by changing work schedule or shift models. For changes beyond a certain volume, the line segments must be re-balanced.		
Expansion flexibility		
An increase in the volume per unit of time can be achieved by increasing the density of workers and the associated re-balancing. If an increase in worker density is no longer possible, the assembly line must be extended.		
Can be successively expanded by installing new cells, even during ongoing production.		

Conclusion Flexibility

In the area of process and routing flexibility, there are no extended advantages in combining the two systems, but this is also only a matter of internal and not market-related flexibility. In the relevant market-related flexibility of product and product-mix, the strengths of both systems are combined very successfully. **The VarioTakt handles the global workload differences by level distributing it the assembly line, while the hybrid assembly handles the local workload peaks.** If the work schedules are exhausted, both systems have to be adapted for an increase in volume; only the Flex Segments of hybrid assembly can be expanded or reduced during ongoing production.

8.2.3.2 Investments

Transport Technologies With the exception of a transport system in which the distances between the assembly objects can be changed, the VarioTakt does not require any additional investments. If such a system is already in use (Sect. 4.6), the only investments are in the one-time adaptation of the transport system control. The investments of Flex Segments are higher than those of comparable line segments. In their case study of an automotive final assembly, Kampker et al. (2021) calculated that the investments of a Flex Segment, which replaces approx. 8 line stations and 2 buffer stations with 3 cells and 10 decoupling buffer stations, are circa 3 million EUR above those of the line assembly.

Floor Space Unfortunately, there are no published simulations of how the floor space requirement changes when a line assembly is transformed into a hybrid assembly. However, it can be assumed that this is lower than the floor space requirement of a pure matrix assembly with the same product range, because in this case the number of work stations increases significantly compared to line assembly (Greschke et al., 2014; Greschke, 2016; Bochmann, 2018; Küpper et al., 2018).

CPS and Control Systems The sensor technology and IT infrastructure of a CPS are investment drivers; regardless of the assembly system, almost all newly installed assemblies increasingly use CPS for control and optimization. The requirements when using Flex Segments are higher than in line assembly, due to the greater control effort, but far lower than in a comparable matrix assembly. Real-time control effort in hybrid assembly only occurs in the Flex Segments and is thus severely limited.

Operating Resources If identical activities are to be carried out in several cells of the Flex Segments, then additional operating resources will be necessary. However, it is to be expected that especially for the installation of large components, the operating resources in the cells of the Flex Segments can be carried out much more easily, since these assembly processes do not have to be carried out in continuous movement, in contrast to line assembly.

Conclusion Investments

Compared to line assembly, hybrid assembly can be expected to involve higher investments (Kampker et al., 2021). In individual cases, requirements for operating equipment in the Flex Segments may be lower, as these do not have to be capable of flow operations. Further research and practical implementation by other pioneers in the application of hybrid assembly are needed to better quantify these investment requirements in comparison with other assembly systems.

8.2.3.3 Operating Costs

Takt Losses One of the major criticisms of an assembly line, in particular mixed-model assemblies of type 3, is the handling of large takt time spreads resulting from a high variance between products and option variants. On the one hand, by using hybrid assembly, local takt time spreads can be managed very well, Kampker et al. (2021) showing a 19% increase in utilization. The VarioTakt, on the other hand, is very good at mastering global takt losses triggered by workload spreads between different products, as we showed in Sect. 4.5.

Utilization Losses It is not the objective of hybrid assembly, and certainly not the objective of line assembly in VarioTakt, to compensate for disruptions, these can lead to a line stop—both systems focus on other priorities. By using the Flex Segments, takt losses can be reduced and model-mix losses avoided. Activities with large takt time spreads are located in the Flex Segments and allow for much better line balancing. Utilization losses due to stops triggered by temporary overload situations (model-mix losses) and utilization of workers (takt losses) triggered by large takt time spreads are now planned and thus can be reduced.

In hybrid assembly, the managerial advantages of flow production, as we described in Sect. 2.1, are retained for most of the assembly process. This promotes continuous improvement of the assembly system, which is an advantage that should not be underestimated.

Similar to matrix assembly, cells in the Flex Segments could employ workers with performance limitations. The performance level of these cells could be reduced for the duration of the staffing by these employees and thus does not reduce the throughput of the entire system.

CPS and Control Systems The combined use of line segments and Flex Segments with predefined series-sequence significantly minimizes the control effort.

Conclusion Operational Costs

In terms of operating costs, both systems, hybrid assembly and VarioTakt, combine their strengths. If local workload peaks or large option variants are processed in the Flex Segments, the VarioTakt compensates for global workload differences between the products.

8.2.3.4 Complexity and Controllability

In hybrid assembly, the complexity of the control is limited to the Flex Segments; the line segments in VarioTakt need not be controlled operationally. The degrees of freedom in series-sequencing are increased by the use of variable takt and by Flex Segments, which should lead to a reduction in complexity in series-sequencing.

8.2.3.5 Ability for Continuous Improvement

We have explained the capabilities of line assembly for continuous improvement in Sect. 7.2.4.5. Assembly lines can be optimized very well with Lean Management methods known so far and regularly applied in practice. Variable takt times and Flex Segments further increase the degrees of freedom here and thus enlarge the solution space. For the optimization of Flex Segments themselves and the decision as to which assembly contents are to be mapped in the Flex Segments, simulations will have to be used to a greater extent, since these interrelationships can no longer be grasped by planners or managers using manual methods.

8.2.4 Conclusion and Outlook Hybrid Assembly

The theoretical and practical development of the hybrid assembly concept has made an important contribution to significantly expanding the product range of type 3 mixed-model assemblies. The reduction of complexity and the limitation of the necessary investments support the implementation. Combined with the VarioTakt, hybrid assembly shows its full strengths in the mastering of a variant-rich type 3 mixed-model assembly line.

The four main advantages of matrix assembly compared to an assembly line in VarioTakt (see Sect. 8.1.3), are reduced by the following two points, because these are now taken over by hybrid assembly:

1. Flexible integration of the heterogeneous workforce
2. Mastery of large takt time spreads of option variants

This is because the individual integration of employees (e.g., due to performance limitations) and the compensation of large option variants are implemented in the Flex Segments. The utilization space of an assembly line is again significantly expanded by hybrid structures, which we show in Fig. 8.4.

The biggest advantage of hybrid assembly, however, could be that it is easier to implement for the existing production management compared to matrix assembly. Not only because of the lower investment costs and control complexity, but also for the reason that this concept only expands the existing experience horizon of an assembly line a bit and does not completely challenge it.

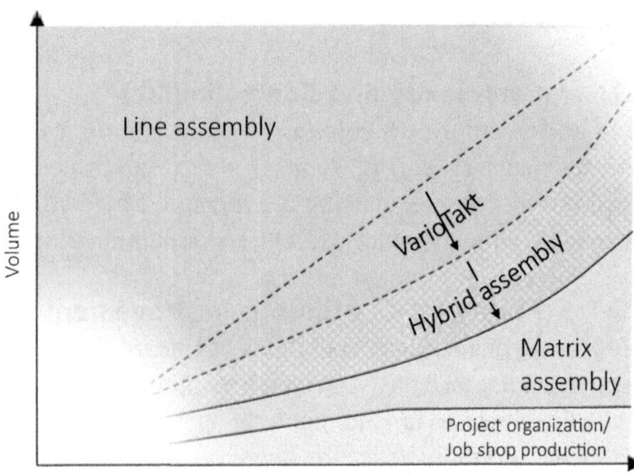

Fig. 8.4 Hybrid assembly and VarioTakt—an extension of the range of use of line assembly

8.3 Cycle Module Assembly at Porsche for Taycan Production

"With the Taycan, we are opening a new chapter. The Taycan is something very special: performance, range, the innovative 800-volt technology for the shortest charging time and the entire vehicle concept are unique," says Porsche Board Member for Production Albrecht Reimold (Porsche, 2020a). The Taycan is Porsche's, initially first, answer for the age of electrification. As the Mission E, the first prototype was presented at the International Motor Show (IAA) in 2015. At the end of the same year, the decision was made for series production and for the Zuffenhausen site in Germany. Initially, this came as a surprise, as the technical conditions at Porsche's Leipzig site would clearly have been better. However, Porsche wanted to manufacture its first thoroughbred electric sports car at the traditional site of its sports cars—the main plant in Zuffenhausen. Thus, the first construction work was started in mid-2016, and production was launched in September 2019, just 4 years after the presentation at the IAA (Porsche, 2020b). In total, more than 700 million EUR were invested in the construction of the new plant. Due to limited space at the site, the Taycan, like the Porsche 911, will be assembled in a multi-story building. In addition, due to the spatial separation of the production buildings, the painted bodies are conveyed to assembly via a 900-meter-long conveyor system (Porsche, 2020b). More than 1500 new employees found jobs in the Taycan production, where its new derivative, the Taycan Cross Turismo, will also be manufactured from 2021.

"Smart, lean, green" are the buzzwords used by Porsche to describe its new production concept for the Taycan (Porsche, 2020b). "Smart" stands for the networked use of CPS systems (see Sect. 7.2.1). The targeted and responsible use of Lean Management methods (see Sects. 2.1 and 3.1) is summarized under "lean" and all ecological aspects with "green." In this production system, too, "the focus should continue to be on people" (Porsche, 2020b) and they should be supported in carrying out their "craft" by means of automation and digitalization.

8.3.1 The Operation of the Cycle Module Assembly

In the assembly structure, Porsche opted for an assembly line in continuous flow operation. For the technical implementation, however, no classic apron conveyor was installed; instead, all car bodies are transported by means of an AGV. Porsche refers to this transport system as a "Flexi Line." Significantly more flexibility in the event of modifications and lower investment costs were the basis for the decision. One of the most important prerequisites for the use of variable takt times would thus be created. Currently, over 200 vehicles are produced per day with a takt time of approximately 4 min.

In the first assembly section, the bodies are still transported by means of C-type hangers, which offers the ergonomic advantage of being able to individually swivel and adjust the height of the assembly objects for work on the underbody of a vehicle. Parallel to this assembly line, the components in the chassis assembly (chassis and battery) are already transported by AGV. After the fully automated marriage, the Taycan vehicles are finally assembled in further stations of the Flexi Line.

Porsche relies on two main concepts to manage variance in Taycan assembly:

- Cycle Module Assembly
- Targeted compensation of work operations

Compensation activities are a concept from PD for MMA. Since this is not intended to be the main focus of our work, we will explain them only briefly below. For option or product differences that lead to local takt time spreads, balancing contents are specifically searched for in case the option variant does not occur. It is obvious that the precedence relationships, and thus the product tree structures, must be specifically influenced during product development.

The method of Cycle Module Assembly is to be located in AD for MMA. It is a method of work organization that allows modified line balancing. The final assembly of the Taycan consists of 8 Cycle Modules. A Cycle Module combines 5 to 7 stations in the assembly of the Taycan, which are processed serially by a worker. In the upper part of Fig. 8.5, we schematically represent a Cycle Module with three takts. The sequence of the assembly activities is specified by a precedence graph. At the end of the Cycle Module, the worker returns to the start of the Cycle Module and starts with a new vehicle. Thus, there are always as many workers in the cycle as there are stations in the Cycle Module. From the worker's point of view, his/her personal takt time is increased by an integer multiple of the takt time of the line. The integer multiple corresponds to the number of stations in the Cycle Module. Looking at it from a different perspective, we could also speak of a Cycle Module consisting of three takts, creating three parallel stations with three times the takt time; we indicate this on the right side in Fig. 8.6. Due to the different size of the Cycle Modules, different takt times per worker can be generated in the line. The takt time per worker could thus be described as variable, but not individually dependent on the product or order, as in the variable takt, but dependent on the size of its Cycle Module. The

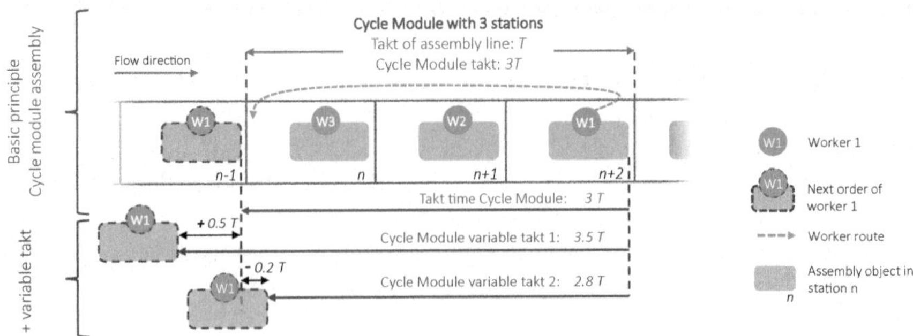

Fig. 8.5 Cycle Module Assembly—three station cycle module with potential adoption of variable takt

combined use of Cycle Module assembly and variable takt (lower area of Fig. 8.5) will be explained later in this chapter.

Prerequisites

The use of workers over several stations requires increased qualification, because the worker must master an assembly content increased by the factor of the stations per Cycle Module. The station-specific use of the equipment (e.g., torque wrench) must be ensured by means of an electronic monitoring system.

8.3.2 Reduction of Takt Losses

Takt losses can be reduced by Cycle Module Assembly. By combining several stations, the takt time spreads, i.e., the takt compensations, are combined for this production area. The takt compensation takes place only once at the end of the Cycle Module and not after each takt/station. The takt time for the individual worker is thus extended by the factor of the number of stations per Cycle Module. Since the absolute amount of work operations remains the same, their relative contribution to the takt time of the Cycle Module is reduced. We have described the positive effects of small operation times (in relation to the takt time) in Sect. 6.3.3.3. We illustrate the positive effects on takt losses with an example in Fig. 8.6. Seven, no longer divisible work operations (WOs) must be balanced among three stations; in our example, this results in a minimum takt time of 3 time units (TU). A lower takt

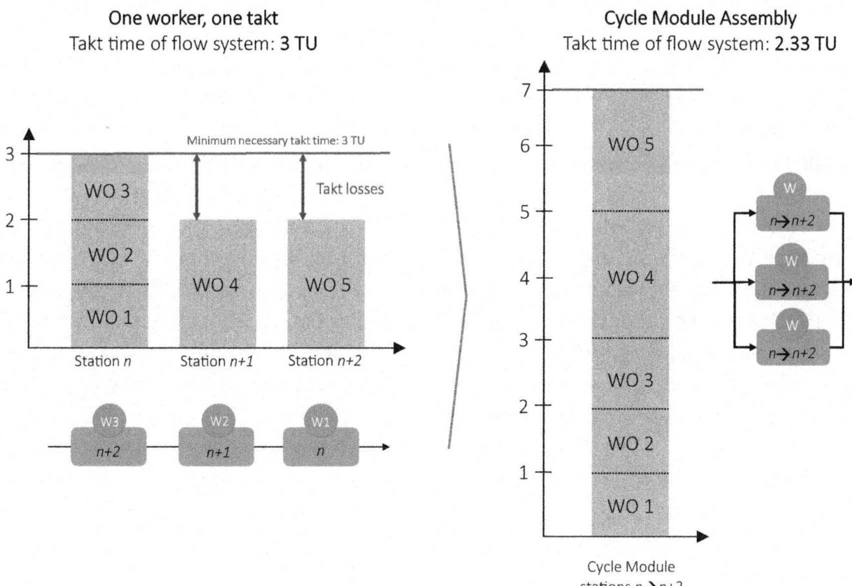

Fig. 8.6 Reduction of takt losses using the example of a three-station Cycle Module

time with three stations is not possible in our example. If these operations are combined in a Cycle Module of three stations, the takt time of the whole system in our example can be reduced to 2.33 TU.

8.3.3 Reduction of Model-Mix Losses

The variance peaks of the individual stations have the possibility to level each other out in the Cycle Module, model-mix losses are reduced. Negatively correlated variants do not have to be located in the same takt/station but can be distributed over several stations in the Cycle Module. In the WATT method, variance is compensated by means of open station boundaries and a drift (see Sect. 2.6.10). Nevertheless, the maximum drift per station must be limited, as this can hinder the following worker in his/her station. Although the average drift per Cycle Module cannot be increased by Cycle Module Assembly, the drift range per station can be designed much more generously, since the current worker him–/herself is his/her successor.

8.3.4 Cycle Module Assembly in Combination with Variable Takt

By selecting an AGV transport system in Taycan assembly, the most important prerequisite for the realization of variable takt is given. Variable takt can also be used purposefully in Cycle Module Assembly to master global variance. If it is desired to give a specific order in the entire assembly an individual takt time per Cycle Module, the distance between two units of the same worker could be adjusted. In the bottom section in Fig. 8.5, we illustrate this concept. In Chap. 4, we describe in detail how variable takt works, where the variable takt is set by adapting the distance to the following unit. In Cycle Module Assembly, the distance must not be set to the following unit on the assembly line, but the distance to the following unit of the respective worker is the quantity to be varied (see Fig. 8.5 lower area). The distance between the other units is irrelevant but must be technically feasible. In this way, the takt time of the individual orders can be increased or decreased by any factor. Limits are set by technical conditions of the assembly line and the product to be assembled, as we describe in Sect. 4.4. This mode of operation further results in the condition that all Cycle Modules of the assembly line in question must consist of the same number of stations. However, the Cycle Modules can be arranged staggered (left side, right side) or overlapping. The thus achieved functioning of the variable takt in the Cycle Module Assembly follows all rules and achieves all advantages of the variable takt, which we describe in this book.

8.4 ARC Assembly at Honda Prachinburi

Certainly, German automakers, driven by their premium products with ever-increasing option variants, are, on the one hand, under particular pressure to master variance in assembly. On the other hand, there is Toyota, the leader in Lean and cost-efficient production, which initially always tried to minimize variance in its products in favor of reduced manufacturing costs. In the meantime, a manufacturing company must master both worlds; otherwise, it will not be successful in the market. A very good example of this path is Honda with its Prachinburi manufacturing plant in Thailand. In 2016, the Prachinburi plant was opened, producing the City, Jazz, and Civic models for Thailand and numerous other Asian countries. Australia and New Zealand are also supplied from this plant.

Honda implemented its ARC concept for the first time at the Prachinburi plant. ARC stands for Assembly Revolution Cell. As Honda's name suggests, this is a hybrid assembly form consisting of a cell and line assembly. However, it is a completely different form than the hybrid assembly structures developed by BMW or RWTH Aachen (Sect. 8.2). In Honda's ARC Assembly, the assembly cell flows in the assembly line. A work system is created that is comparable to the Cycle Module Assembly of Porsche's Taycan production. In MMAD, ARC Assembly is thus a further implementation of elements of AD for MMA.

The basic structure of the ARC Assembly is initially similar to that of a classic plate/apron conveyor. However, a conveyor element consists of a round and a rectangular element, which are movably connected to each other (see Figs. 8.7 and 8.9). Thus, not only linear but also circular movements of the line are possible. The vehicle is positioned on the rectangular element. On the ARC unit, kitting carts carry all the materials for an assembly section of 6 to 10 stations (see the lower part of Fig. 8.9).

The entire line consists of 50 ARC platforms, which are arranged in several U-shaped segments to form an assembly line. Each U-shaped segment forms an assembly section in which a team of four workers performs the assembly and flows

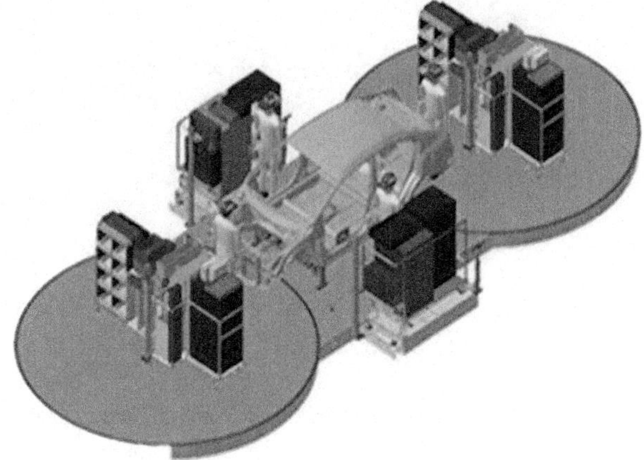

Fig. 8.7 ARC element (Source: Honda Motor Company)

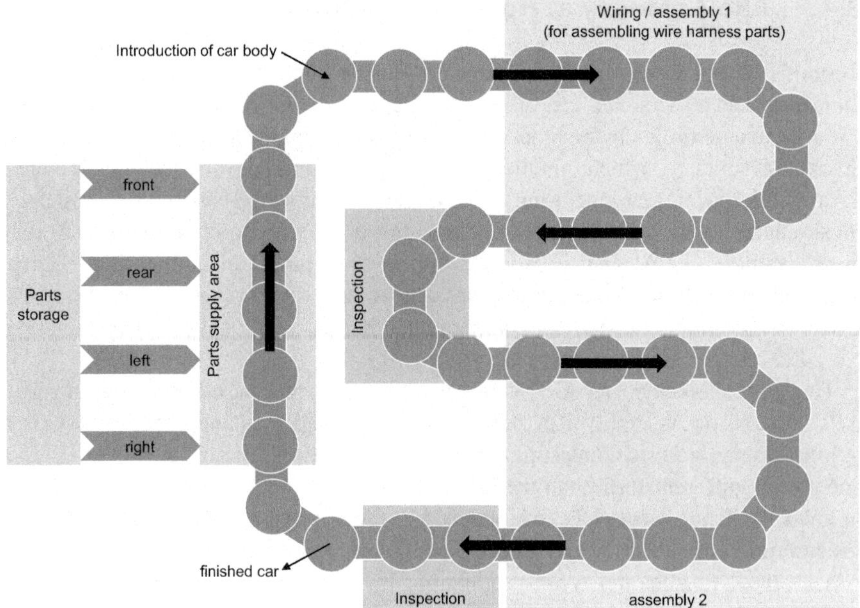

Fig. 8.8 ARC Assembly (own illustration, based on Nikkei Inc.)

Fig. 8.9 U-shaped line of ARC Assembly (Source: Honda Motor Company)

with the assembly object within the segment (Fig. 8.8). Workers and materials are assigned to the four vehicle sides: front, rear, left, and right. Since the beginning and end of each assembly segment are very close together due to the U-shaped layout (Fig. 8.9), the paths of the worker teams are very short when switching to a new order.

8.4.1 Evaluation of the ARC Assembly

- Honda reports that the ARC concept can achieve a 10% increase in efficiency. The main reasons for this are due to the shortened walking distances for the workers, as supplies are positioned a short distance away on the ARC element.
- As a further advantage, Honda states that the assembly can be easily extended by its U-shaped lines, which are reminiscent of the finger structure of the BMW plant in Leipzig (Sect. 2.3), by adding additional ARC elements to extend the segments.
- Since the entire conveyor system, by means of a chain drive, is applied to the floor and not into the ground, the installation costs would be lower than for the classic apron conveyor.
- In ARC Assembly, the same advantages are achieved as in Cycle Module Assembly (Sect. 8.2). Variance is levelled and balanced over several stations. However, the term "stations" is not appropriate in ARC Assembly because the workers move with the ARC cell in the assembly segment.
- Due to 100% kitting of all parts, minimal space is occupied by material on the line, and a high number of products or option variants can be assembled in this way.

Challenges certainly include:

- The maintenance and servicing of the complex conveyor system.
- The 100% kitting of all parts of a vehicle is a very complex process. It must be possible to carry out the process very efficiently and, above all, without errors.
- Although teamwork across several stations is intended to generate efficiency gains by reducing takt losses, the qualification effort required of the workers is many times greater than in the classic single-takt assembly line.
- The special design of the ARC Assembly eliminates some of the advantages of a classic flowing assembly as described in Sect. 2.1. For example, stopping the assembly line in case of problems can be avoided by calling additional workers into the line. Together with the team, they then have time until the order reaches the end of the U-shaped segment. The impact of problems is thus mitigated and, under certain circumstances, does not lead to a lasting elimination of the real root cause.

8.4.2 ARC Assembly in Combination with Variable Takt

Due to the fixed mechanical connection of the ARC cells, no variable takt is possible in the Honda system. If the basic idea of ARC Assembly, including its layout, is transferred to an AGV system in which both the orders and the material carts are transported continuously on individual AGVs, a variable takt with all its advantages could also be implemented in ARC Assembly. The spacing of the orders would have to be adjusted as we have already described in the Cycle Module Assembly in Sect. 8.3.4. Another example of how many different assembly concepts the variable takt can be combined with.

8.5 Load-Oriented Series-Sequencing at Mercedes-Benz Factory 56

With its "Factory 56," which opened in 2020, carmaker Mercedes-Benz also wanted to present its new model for future assemblies, whose core competence is to lie in the mastery of variance. Thus, at the start of the new Factory 56, only the S-Class is produced initially, followed by Mercedes's Maybach S-Class and the EQS, a fully electric vehicle. According to Mercedes-Benz wishes, it should be possible to integrate further models, from compact cars to SUVs, within a very short time and during ongoing production. This proof must be provided in the next few years though. Should Mercedes-Benz succeed in assembling the S-Class with combustion engine and the EQS electric vehicle together on a single assembly line in continuous flow operation, a new standard would initially be set in automotive production.

A high level of integration of current I4.0 technologies, such as 5G networks, a wide variety of digital end devices, and a holistic CPS form the technological basis, but should not be the focus of our considerations.

High flexibility is to be generated above all by the assembly system. In the case of Factory 56, this is a classic assembly line in a fixed takt. Similar to the Taycan assembly of the competitor Porsche AGVs are mainly used instead of an apron conveyor—400 in total. Unfortunately, the AGVs are not yet used to generate a variable takt via variable distances, variable launching intervals respectively. The main purpose of the AGVs is to maintain flexibility in the routing so that it can be adjusted as needed without major technical intervention. In addition, the assembly line is designed in such a way that, as far as possible, no fixed technical points are created or the existing fixed technical points are concentrated in order to be able to design the rest of the assembly line very flexibly. Interestingly, this goes hand in hand with a decrease in the automation of assembly processes. Several times in this book, we have already discussed the fact that smaller, globally distributed assembly lines will emerge in the future, which will have to handle significantly more variance. One consequence is that the focus will be on automating all non-value-adding processes, especially in logistics. It is not the assembly processes themselves that are the challenge; these will be carried out by qualified employees in a high-quality and very flexible manner without major investments. Mastering variance and assembling with a high variety of products and parts on a small footprint will become the dominant objective.

It is interesting for the focus of our considerations that Mercedes-Benz does not use any model-mix rules, nor car sequencing, for series-sequence determination in its Factory 56 for the first time. Instead, it completely chooses the overload-oriented approach of model-mix sequencing, as we describe in Sect. 3.2.3. Thus, up to 60% overbalancing ($\beta = 0.6$!) *can* be mastered. Assembly line balancing is largely automated, and the main task of the assembly planners is to establish the basics and basic data for (partially) automated balancing. Series-sequencing is then fully automated, with the aim of minimizing the use of additional flexible workers.

We do not have any details on the exact technical design of the entire conveyor system at Factory 56. However, for the part of the conveyor system in which AGVs are used, it can be assumed that a variable takt could be implemented without major changes. If, as intended, additional products are added to the assembly line whose total assembly workload differs from that of the existing products, the degrees of freedom in model-mix sequencing will certainly decrease. The use of variable takt would significantly increase these degrees of freedom again.

8.6 Automated Sequencing and Series-Sequence Planning with Variable Takt and VarioTakt: An Experiment

Increasingly, different software solutions for automated balancing and series-sequencing are available on the market. We use one of these applications for a benchmark study, an experiment between fixed and variable takt—with very interesting results. For this purpose, we selected a software that is well established on the market and is also used by large automobile manufacturers, among others. We are aware that such applications initially represent a "black box," since no detailed knowledge of the optimization algorithms and other software components are public, but we were able to reproduce the results obtained and classify them as highly plausible.

After entering all work operations with operation times, their precedence relationships, other restrictions, and a production program with configured orders for a planning period, the software program enables a supported or fully automated optimization of the line balancing according to different objectives. Furthermore, this application enables workload-oriented series-sequencing, as we describe it in Sects. 3.2.3 and 3.2.4 as model-mix sequencing.

Unfortunately, to this date, no software application that we know of has the ability to properly map variable takt or the VarioTakt. With the help of the selected software though, we can show in the last section of our work how variable takt can be used in type 3 mixed-model assembly with two different products in order to reduce takt losses and increase the efficiency of the assembly line. With the help of automated line balancing, it is also possible with very little effort to create the "best" line balancing for a different model-mix in each case, which we show in Table 8.1 and Fig. 8.10. In our following considerations, "best" line balancing stands for the line balancing with the highest balancing efficiency and compliance with all restrictions.

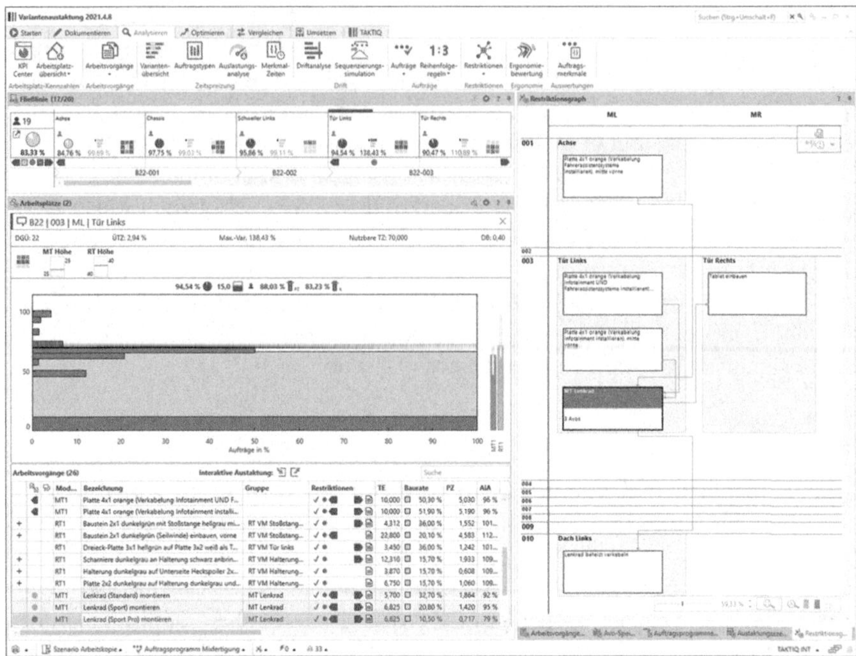

Fig. 8.10 Software work screen—automated line balancing

8.6.1 Experimental Design in Automated Line Balancing

For our experiment, we have set up an assembly line with 20 workstations $N = 20$ and $W = 22$ workers in the line balancing software. In the following, we will use the same designations for workers and workstations. The stations are based on an automotive assembly line in terms of structure, work content, and designation. Two different products are assembled on our experimental assembly line: Monster Truck and Rocket Truck (Fig. 8.11). The two vehicles differ significantly in their basic product architecture and thus also in the BOM structure, so that we can define them as two independent products $P_{MonsterTruck} = P_{MT}$ and $P_{RocketTruck} = P_{RT}$ according to our definitions from Sect. 2.4. Customers of our two products have the option to choose from 119 different option variants. All operation times ot_l are defined, and the complete precedence relationships are documented in the line balancing software. In total, there are 395 work operations (WOs) to be balanced.

For the determination of the probability of occurrence P_l of the ot_l (see also Sect. 2.6.8), a production volume of 1000 units is utilized. In this program, initial P_{MT} takes about 70% of the volume and P_{RT} about 30%. For all following considerations in this chapter we set the allowance times to zero: $e_a = e_{fa} = e_{pa} = 0$.

Figure 8.11 shows our line balancing situation graphically. The total weighted operation times of our two products differ by 31.8%. The curve of the increasingly

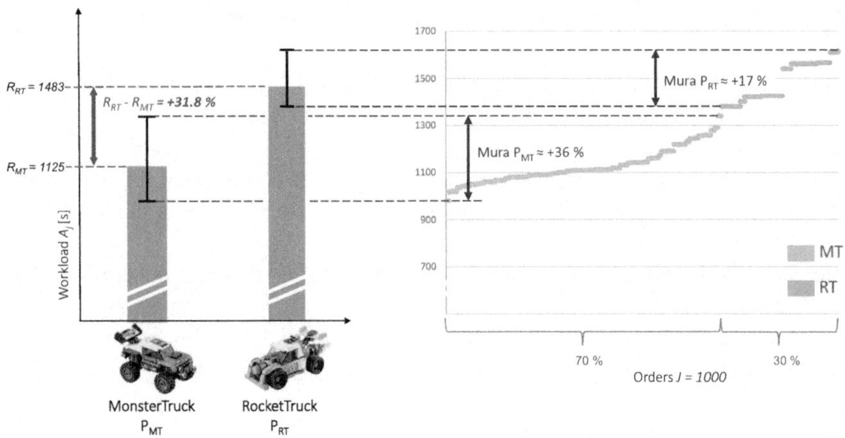

Fig. 8.11 T-Bar and curve of the increasingly sorted A_j of the 1000 orders of Monster Truck and Rocket Truck

sorted workloads of both products does not overlap in this case, i.e., $A_{MT\_max} < A_{RT\_min}$. Mura in P_{MT} is significantly larger compared to Mura in P_{RT}, option variants have a larger impact in P_{MT}. The impact of this difference will become apparent as our simulation progresses.

Restrictions

In addition to classic precedence relationships, due to the precedence graph of the two products, further restrictions can be applied in an automated line balancing:

- *Group restrictions*: Defined WOs must be performed together, by one worker. Work operations must be technically conditioned (e.g., cleaning and gluing) or assembled with the same operating resources.
- *Exclusions*: Certain WOs may not be performed together, i.e., by one worker. For example, activities such as greasing and gluing must be separated, or the implementation of the dual control principle to safeguard critical assembly operations is desired.
- *Spacing restrictions*: Activities must be spaced a maximum or minimum number of stations apart, e.g., gluing must occur no later than two stations after cleaning. The primer must evaporate at least three stations before gluing can take place.
- *Position restrictions*: The mounting object must be in a certain position, e.g., height.
- *Operating resources*: WOs can only be assigned to stations with specific operating resources.

Following an actual assembly, all stations and WOs in our experiment are assigned different characteristics or restrictions.

8.6.2 Main and Secondary Conditions of the Simulation

In automated balancing, after entering the takt times T_i different objectives can be optimized by an automated shifting of the WOs and under consideration of all precedence relationships or restrictions. In the following experiment, simulation respectively, we always use the WATT balancing method (Sect. 2.6) in addition to the variable takt and thus the VarioTakt. We use T_{Vario} for the takt time in the VarioTakt.

Objectives: Balancing Utilization or Average Utilization of the Assembly Line Our objective to be maximized is balancing efficiency and thus the average utilization of the assembly line. In Sect. 2.6.11, we have already defined the formula for the calculation in the fixed takt. In the following, we extend this for the variable takt. The sum of the weighted workload times per worker and takt time group is $R_{w,i}$, this is divided by the total capacity of this takt time group $W*T_i$ and then weighted by the factor p_i. p_i stands for the weighting of the number of orders in the takt time group i in relation to the total number of orders I.

$$E_{AL} = \frac{C_b}{C_v} \cdot 100[\%] = \sum_{i=1}^{I} \left(\frac{\sum_{w=1}^{W} R_{w,i}}{W \cdot T_i} \cdot p_i \right) \cdot 100[\%]$$

E_{AL} Balancing efficiency of the assembly line
C_b Actual capacity required
C_v Provided capacity
i Index takt time groups, $i = 1\ldots I$
w Index worker, $w = 1\ldots W$
$R_{w,i}$ Weighted workload of worker w in takt time group i
T_i Takt time of takt time group i
p_i Proportion of orders in takt time group i in relation to the total number of orders I

Since we want to show the advantage of the variable takt over the fixed takt, we calculate the difference of $E_{AL\_Vario} - E_{AL\_Fix}$ for different mix ratios of our two products. Table 8.1 presents these results.

Constraint 1 Maximum utilization of each worker

We set as a termination criterion for all subsequent balancings of the experiment that the weighted load of each worker must not be greater than the weighted takt. In short: no worker may receive an expected workload greater than 100% on average.

$$\sum_{i=1}^{I} (\max_w \; R_{w,i} \cdot p_i) \leq \sum_{i=1}^{I} (T_i \cdot p_i).$$

Constraint 2 Limitation of the load of the maximum order variant

For all following considerations, we set as a second termination criterion the condition that for no worker the load of a specific order variant may be greater than 115% of the takt time T_i (limitation of the WATT method).

$$\max_w [a_{w,j}] \leq (1 + 0.15) * T_j;\ j = 1 \ldots J$$

T_j	Takt time of order j
$a_{w,j}$	Assembly content of order j at worker w

Execution This experiment does not use OR (Operations Research) optimization in the classic sense, but enumerates over the discrete solution space. An optimization possibility is not given in the software used. We reduced the takt times for each model-mix by one-second increments, thus continuously increasing the line utilization until one of the two constraints (for the currently selected takt times) was no longer fulfilled.

The runtime for automated sequencing was limited to 3 min on a notebook (Intel® Core™ i5-8350U CPU @ 1.70GHz processor; 7.9GB RAM; Windows 10 operating system; manufacturer DELL). *(No significant improvements could be achieved after 3 min of runtime).*

8.6.3 Results

Results in Fixed Takt

Considering the constraints, a minimum takt time $T_{min\_Fix\_7/3} = 70$ s could be achieved with the set production program (Fig. 8.11). $E_{AL\_max\_Fix\_7/3}$ is 80.4% in this line balancing.

$T_{min\_Fix\_7/3}$	Minimum fixed takt time for model-mix MT/RT = 7/3
$E_{AL\_max\_Fix\_7/3}$	Maximum possible balancing utilization (=minimum takt time) in the fixed takt for the MT/RT model-mix = 7/3

Results in the VarioTakt

The same production program was automatically balanced in the VarioTakt, achieving $T_{min\_Vario\_MT\_7/3} = 61$ s and $T_{min\_Vario\_RT\_7/3} = 73$ s. $E_{AL\_max\_Vario\_7/3}$ amounts to 86.7% in this balancing variant. This results in an increase in line efficiency of 6.3% when using the VarioTakt compared to the fixed takt.

$T_{min\_Vario\_MT\_7/3}$	Minimum variable takt time of takt time group MT for model-mix MT/RT = 7/3
$T_{min\_Vario\_RT\_7/3}$	Minimum variable takt time of takt time group RT for model-mix MT/RT = 7/3
$E_{AL\_max\_Vario\_7/3}$	Maximum possible balancing utilization (=minimum takt times) in the VarioTakt for the model-mix MT/RT = 7/3

Modifications in the Model-Mix

In a further step, we can demonstrate the ability of the VarioTakt to independently balance the workload on the assembly line when the model-mix changes. We have already described this effect in Sects. 3.2.4 and 4.5.3. For this we change the model-mix of products P_{MT} and P_{RT} in the production program, the results are shown in Table 8.1 as well as Figs. 8.12 and 8.13.

For the four following scenarios, the automated line balancing is performed in the software, and the results are visualized in Table 8.1.

1. For the fixed takt, for each product-mix variant of P_{MT} and $P_{RT,}$ the minimum possible fixed takt T_{min_Fix} (and thus the maximum balancing efficiency $E_{AL\_max\_Fix}$) was determined by means of the automated line balancing—a modified line balancing was performed for each product-mix.
2. The minimum T_i for each product-mix variant of P_{MT} and P_{RT} was also determined for the VarioTakt. For the determination of the minimum possible VarioTakt for each mix, a modified line balancing was performed.
3. With $T_{Static\_Fix} = 72$ s, a fixed takt was selected that fulfills both constraints for a very large range of the MT/RT model-mix (from 10/0 to 2/8). Thus, this fixed takt can be used to operate the assembly line in these model-mix variants without the need for a re-balancing. The balancing efficiency achieved in this way for each model-mix is shown in Table 8.1.
4. Also, in the VarioTakt, the objective was to be able to cover a large range of the model-mix without re-balancing, for this reason, $T_{Static\_Vario\_MT} = 62$ s and $T_{Static\_Vario\_RT} = 73$ s was set. This also made it possible to cover a wide range in the model-mix from 10/0 to 2/8.

T_{min_Fix}	Minimum fixed takt time for the respective model-mix scenario
$E_{AL\_max\_Fix}$	Maximum balancing utilization in the fixed takt for the respective model-mix scenario
T_{Static_Fix}	Static fixed takt (in our experiment = 72 s) used across all model-mix scenarios
$T_{Static\_Vario\_MT}$	Static VarioTakt for takt time group MT (in our experiment = 62 s) that is used across all model-mix scenarios
$T_{Static\_Vario\_RT}$	Static VarioTakt for takt time group RT (in our experiment = 73 s) used across all model-mix scenarios

Figure 8.12 shows the minimum possible takt times for each model-mix scenario. It is striking in our results that, especially in the model-mix scenarios with a high weighting of the "smaller" product P_{MT}, the VarioTakt is most clearly superior to the fixed takt, due to a better balancing efficiency (Fig. 8.13). It can also be seen very clearly that in each model-mix scenario, the VarioTakt time of the "larger" product P_{RT} is only slightly above the respective fixed takt, but the VarioTakt time of P_{MT} is clearly below it (Fig. 8.12). It can be said that even with a few units of the "large" product $P_{RT,}$ the fixed takt is driven upward (by constraint 2). The minimum VarioTakt time of RT as well, but due to its lower weighting, the average VarioTakt $\emptyset T_{min\_Vario}$ is only slightly lifted (Fig. 8.12).

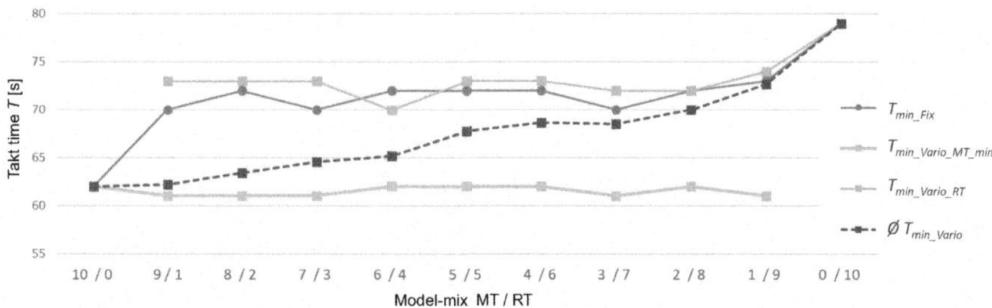

Fig. 8.12 Minimum takt times for different model-mixes

In the middle range of the mix scenarios between 9/1 and 2/8, all takt times are almost constant despite changing total workload, i.e., the workload increases in this direction (Fig. 8.12)! In the model-mix scenario 10/0, there is no value for $T_{min\_Vario\_RT}$ because there is no product of the takt time group P_{RT} in this mix scenario. The same applies in the mix scenario 0/10 for P_{MT}.

$T_{min\_Vario\_MT}$	Minimum VarioTakt time of the takt time group MT for the respective model-mix scenario
$T_{min\_Vario\_RT}$	Minimum VarioTakt time of the takt time group RT for the respective model-mix scenario
$\emptyset T_{min\_Vario}$	Weighted (in ratio of MT and RT) minimum VarioTakt for the respective model-mix scenario (blue dashed line in Fig. 8.12)

Table 8.1 and Fig. 8.13 show that the VarioTakt achieves a better balancing efficiency than the fixed takt for every model-mix scenario. In the model-mix scenarios 10/0 and 0/10, there is no difference in E_{AL}, since in these cases only one product is on the assembly line. The maximum value of the balancing efficiency is reached with the fixed takt at the model-mix scenario $P_{MT}/P_{RT} = 3/7$, and with the VarioTakt at $P_{MT}/P_{RT} = 2/8$ (see Table 8.1).

The significant drop in E_{AL}, or lower values of E_{AL}, in the model-mix variants in which P_{MT} is weighted higher is striking. In our simulation, constraint 2, *maximum variant* $< 1.15^*T_i$, turned out to be a termination criterion in all mix scenarios (except 0/10 and 10/0). Figure 8.11 marks the larger Mura of P_{MT}, these stronger variations due to option differences lead to the generally lower E_{AL} with high weighting of P_{MT}. For this reason, the balancing efficiency E_{AL} is also greater for $P_{MT}/P_{RT} = 0/10$ than for $P_{MT}/P_{RT} = 10/0$.

$E_{AL\_max\_Fix}$	Maximum balancing utilization in the fixed takt for the respective model-mix scenario
$E_{AL\_max\_Vario}$	Maximum balancing utilization in the VarioTakt for the respective model-mix scenario
$E_{AL\_Static\_Fix}$	Balancing efficiency at a static fixed takt (in our experiment = 72 s) used across all model-mix scenarios
$E_{AL\_Static\_Vario}$	Balancing efficiency at same VarioTakt (in our experiment $T_{MT} = 62$ s and $T_{RT} = 73$ s) used across all model-mix scenarios

Table 8.1 Comparison of productivity and minimum takt times of the fixed takt and the VarioTakt with a changed model-mix in the production program

		Model-mix scenarios P_{MT}/P_{RT}	10/0	9/1	8/2	7/3	6/4	5/5	4/6	3/7	2/8	1/9	0/10
1.	Fixed takt (each own line balance)	T_{min_Fix} [s]	62	70	72	*70*	**72**	72	72	70	72	73	79
		$E_{AL\_max\_Fix}$ (%)	82.5	75.4	75.6	*80.4*	80.9	83.0	84.6	89.3	89.1	90.1	85.3
2.	VarioTakt (each own line balance)	$T_{min\_Vario\_MT}$ [s]	62	61	61	*61*	62	62	**62**	61	62	61	–
		$T_{min\_Vario\_RT}$ [s]	–	73	73	*73*	70	73	**73**	72	72	74	79
		$E_{AL\_max\_Vario}$ [%]	82.5	84.9	85.8	*86.7*	88.5	88.2	88.8	91.1	91.6	90.5	85.3
$E_{AL\_max\_Vario} - E_{AL\_max\_Fix}$ (%)			**0**	**9.5**	**10.2**	**6.3**	**7.6**	**5.2**	**4.2**	**1.8**	**2.5**	**0.4**	**0**
3.	Fixed takt	T_{Static_Fix} (s)	**72**									*Takt times violate constraints, therefore no solution exists.*	
		$E_{AL\_Static\_Fix}$ (%)	71.1	73.3	75.6	77.8	80.1	83.0	84.6	86.8	89.1		
4.	VarioTakt	$T_{Static\_Vario\_MT}$ (s)	**62**										
		$T_{Static\_Vario\_RT}$ (s)	**73**										
		$E_{AL\_Static\_Vario}$ (%)	82.5	83.6	84.7	85.8	86.9	88.2	88.8	89.7	90.6		
$E_{AL\_Static\_Vario} - E_{AL\_Static\_Fix}$ (%)			11.5	10.3	9.1	8.0	6.8	5.2	4.2	2.9	1.5		

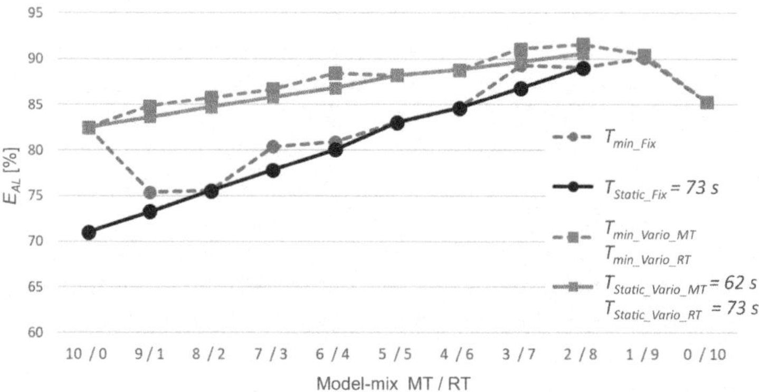

Fig. 8.13 Balancing efficiency with different model-mixes

The lower part of Table 8.1 (balancing variants 3 and 4) shows the comparison between the static fixed takt and the static VarioTakt across all mix scenarios. In this comparison, the VarioTakt plays out its advantages: Over a very large range of the model-mix, no re-balancing is necessary, and the balancing efficiency always reaches higher values than in the fixed takt.

The dashed lines in Fig. 8.13 show the maximum balancing efficiency that can be achieved in the fixed and VarioTakt if a new minimum takt time (T_{min_Fix}; $T_{min\_Vario\_MT}$ and $T_{min\_Vario\_RT}$) is selected for each model-mix scenario. In practice, however, this would mean a new additional line balancing for each model-mix scenario, which represents a considerable (time) effort. However, if the takt times are static (thus no new line balancing is necessary), i.e., $T_{Static\_Fix} = 73$ s, $T_{Static\_Vario\_MT} = 62$ s, and $T_{Static\_Vario\_MT} = 73$ s, the respective balancing efficiency is only slightly lower. In each scenario, the VarioTakt is superior to the fixed takt due to better balancing efficiency. Only in the model-mix scenarios 1/9 and 0/10 can the static takt times not be used because constraints 1 or 2 are no longer fulfilled.

In our last simulation the VarioTakt again demonstrates its ability to adapt independently to a fluctuating model-mix. Even if we keep the VarioTakt static (no re-balancing for mix modifications), it achieves better balancing utilization values than the fixed takt T_{min_Fix} which is adjusted for every mix (Fig. 8.13).

8.6.4 Variable Takt in Workload-Oriented Series-Sequencing

In this last substantive section on advanced concepts for mastering variance in assemblies, we would like to outline another possible use for variable takt.

So far, we have used the variable takt in the context of line balancing in all the considerations in this book. Products have been divided into takt time groups. All products in a takt time group are balanced with the same variable takt time, a more levelled workload on the workers and higher productivity are usually the result. In workload-oriented series-sequencing, orders are sequenced in such a way that the overloads of the different orders cancel each other out at as many stations as possible (Sect. 3.2.3). Ultimately, in workload-oriented series-sequencing, worker drift, due to WATT balancing, is minimized across all stations. Major automotive manufacturers are increasingly using software to perform workload-oriented series-sequencing. Mercedes-Benz, with its new model factory Factory 56, relies on software solutions developed in-house, while other manufacturers, such as BMW, use solutions from third-party software vendors.

If an assembly line already masters and uses variable takt, it is technically capable of **processing each individual order with an individual takt time (respectively distance) in the assembly line**. The variable takt time of the products in the takt time group mainly compensates for the differences in utilization that are due to the basic structure of the individual products (global variance). Differences in utilization losses, in particular model-mix losses, of option variants are compensated for in the VarioTakt by the WATT method and mastered by drifting of workers (local variance). It is precisely this drifting that is levelled out in workload-oriented series-sequencing.

In other words: In a takt time group, all orders oscillate around the takt time of the takt time group (see workload equilibrium in Sect. 5.3); ideally, overload and underload of the individual orders of a takt time group are perfectly levelled (see also VTGA in Sect. 5.3). This condition initially applies to the planning phase of the assembly line, on the basis of which the line balancing is carried out. However, this involves the consideration of a period of time, not a point in time. What if, due to existing customer orders for the day (point in time), no balance of overload and underload in the respective takt time group can be achieved by series-sequencing? The demand for orders with many and few option items is not balanced—could not *each order* then be balanced in its *own individual takt time?*

In practice, it is not possible to re-balance each individual order with an individual takt time in order to master option differences. In the case of fully automated line balancing, the planning effort could be kept very low. In the case of individual line balancing for each order, the workers on the assembly line would have to be qualified for each (thus endless) line balancing variants, which is not possible from an organizational and cognitive point of view (too high complexity for the workers). Line balancing is created in the planning phase of an assembly line and is no longer changed in the control phase, i.e., series-sequence planning.

What if it were possible to give each individual order a "small" bonus or deduction in its variable takt time, i.e., during workload-oriented series-sequencing which affects its individual launching distance D_j instead of changing the actual line balancing. In the end, it does not depend on the respective individual order, but on the current line situation! A single order with a particularly large number of option items does not yet present the assembly line with a major challenge; the WATT can absorb this. However, if a large number of orders with many option variants follow (because these are currently in demand), the workers drift more strongly due to the WATT. Identical orders could thus receive different takt times when using variable takt in the series-sequencing phase, but at a different time respectively at a different workload situation in the assembly line. The line balancing of these orders is not changed and remains identical, thus also the demands on the affected workers in the assembly line.

A selective increase in the takt time of individual orders could increase the degrees of freedom in series-sequence planning or reduce the use of flexible workers. A selective reduction in the takt times of individual orders, in the event of a temporary underutilization, would have the same levelling effect. The use of variable takt in sequencing could thus significantly increase the degrees of freedom for a true build-to-order.

In summary, it could be said that variable takt takes over the levelling task for global variance (Sect. 2.4.2) between the different products in the phase of line balancing. If variable takt is also used in the control phase, in workload-oriented series-sequencing, local variance can also be levelled by option variants and the currently prevailing model-mix.

References

Bochmann, L. S. (2018). *Entwicklung und Bewertung eines flexiblen und dezentral gesteuerten Fertigungssystems für variantenreiche Produkte* (185 p.). ETH Zurich. https://doi.org/10.3929/ethz-b-000238547

Greschke, P. (2016). *Matrix-Produktion als Konzept einer taktunabhängigen Fließfertigung* (180 p). Vulkan Verlag.

Greschke, P., Schönemann, M., Thiede, S., & Herrmann, C. (2014). Matrix structures for high volumes and flexibility in production systems. *Procedia CIRP, 17*, 160–165. https://doi.org/10.1016/j.procir.2014.02.040

Hottenrott, A., & Grunow, M. (2019). Flexible layouts for the mixed-model assembly of heterogeneous vehicles. *OR Spectrum, 41*, 943–979. https://doi.org/10.1007/s00291-019-00556-x

Kampker, A., Kawollek, S., Fluchs, S., & Marquard, F. (2019). Einfluss der Variantenvielfalt auf die automobile Endmontage. *ZEW, 114*, 7–8. https://doi.org/10.3139/104.112097

Kampker, A., Kawollek, S., Marquard, F., & Krummhaar, M. (2021). Potential of hybrid assembly structures in automotive industry. *Aachen University* Library, 10 p. https://ssrn.com/abstract=3848756

Kern, W., Rusitschka, F., Kopytynski, W., Keckl, S., & Bauernhansl, T. (2015). Alternatives to assembly line production in the automotive industry. *23rd International Conference on Production Research* (9 p.). http://publica.fraunhofer.de/eprints/urn_nbn_de_0011-n-3798198.pdf

Küpper, D., Sieben, C., Kuhlmann, K., & Ahmad, J. (2018). Will flexible-cell manufacturing revolutionize Carmaking? *The Boston Consulting Group Online*, October 18. Accessed April 28, 2021, from https://www.bcg.com/de-de/publications/2018/flexible-cell-manufacturing-revolutionize-carmaking

Porsche. (2020a). Albrecht Reimold: "Zuffenhausen is the cradle of our sports cars". *Porsche Newsroom.* Accessed March 07, 2021, from https://newsroom.porsche.com/de/unternehmen/porsche-taycan-zuffenhausen-zero-impact-factory-fabrik-der-zukunft-produktion-4-0-elektromobilitaet-18489/interview-albrecht-reimold-18597.html

Porsche. (2020b). The factory of the future – Smart, lean and green. *Porsche Newsroom.* Accessed March 07, 2021, from https://newsroom.porsche.com/de/unternehmen/porsche-taycan-zuffenhausen-zero-impact-factory-fabrik-der-zukunft-produktion-4-0-elektromobilitaet-18489/fabrik-der-zukunft-smart-lean-green-18594.html

Sethi, A. K., & Sethi, S. P. (1990). Flexibility in manufacturing: A survey. *International Journal of Flexible Manufacturing Systems, 2*, 289–328. https://doi.org/10.1007/BF00186471

Summary and Outlook 9

With this short chapter, in the style of a management summary, we would like to generate motivation for implementation in a condensed form. We provide the most important core arguments of all chapters on the introduction of variable takt and the VarioTakt. We place the variable takt and our further elaborations in our Mixed-Model Assembly Design (MMAD) and give a short outlook on possible future developments.

9.1 Variable Takt Is Simple, Successful, and Necessary

It is decisive! **Variance has to be mastered, not reduced, in a flow assembly**—variable takt and the VarioTakt are excellent production management tools for achieving this goal. With increasing mass customization and emerging mass individualization, as we describe in Sect. 1.2.1, mastering variance through variable takt is a crucial competitive advantage of manufacturing companies in developed industrialized countries. Combined with the effect of a resilient, globally distributed and customer-centric assembly structure, there are no alternatives to being able to successfully manufacture a wide variety of products in one assembly in the future without variable takt.

At the same time, products individualized according to customers must not be standardized; this competitive advantage must be maintained. Customer variance in assembly must be mastered, not suppressed or "standardized away." It is not the product itself or the production program that must be standardized in order to level the (assembly) processes! The assembly system itself must be able to master the variance in the most diverse products, to smooth out the strain on people and the system.

It is important! The **advantages of continuous flow production must be maintained** (Sect. 2.1)—despite increasing variance in and between products of an assembly. Continuously flowing processes force us to perfection; even the smallest problems come to the surface (Liker, 2021) and have to be solved once

P. Bebersdorf, A. Huchzermeier, *Variable Takt Principle*, Management for Professionals, https://doi.org/10.1007/978-3-030-87170-3_9

and for all. Overproduction, and thus waste of various kinds, is avoided, the positive effects of the line pressure of a flowing assembly line can be used.

It is equalizing! On the one hand, as promising as assembly according to the build-to-order principle is, the fluctuations in the workload are a challenge—for the individual person, as well as the processes and thus the results of the organization. Assembling a wide variety of products, or entire product portfolios, on an assembly line in a fixed takt cannot represent a continuous flow of workload. Using variable takt, on the other hand, creates **a steady flow of workload for all employees on the assembly line**. Variable takt can not only show its advantage in balancing, but also in sequencing, where an additional leveling effect can be achieved (Chap. 3 and Sect. 8.6.4).

It is simple! Variable takt and the VarioTakt can be implemented without a complex information technology infrastructure, without self-learning algorithms, without automation, without robotics or artificial intelligence. If a high variance of products is to be mapped in a fixed takt, this creates a high level of complexity—it is a contradiction in itself! **Variable takt prevents this complexity from arising in the first place**. Each order in the assembly line gets the assembly time it needs—no more shifting, drifting, or temporary additional personnel is required. The only major hurdle to successful implementation often remains management, with its entrenched thought patterns and risk aversion to new ideas. Use this book to show: It works!

It is successful! **Nothing increases the productivity of an assembly line with different products more than the VarioTakt.** If the demanded product mix changes, the VarioTakt keeps the productivity of the assembly line much more constant than all other common assembly systems. The need for mix-related re-balancing is significantly reduced. The output in terms of workload, and in most cases also in terms of sales, becomes significantly less dependent on the mix of products. Investments for the integration of new products into the assembly are reduced. Line sequence restrictions due to differences in product utilization: eliminated with the VarioTakt! A build-direct-to-order becomes more realistic. All these advantages are described in detail in Sect. 4.5.

9.2 Mixed-Model Assembly Design: The MMAD-Model

We organize our approaches to mastering variance in assemblies using Mixed-Model Assembly Design (MMAD), which we define in many places in our work.

In Chap. 6, we outline approaches of Design-for-Takt in the Product Design for Mixed-Model Assembly (PD for MMA) in order to create the conditions in the *design phase* of the products to be able to better master variance in assembly. We thus enlarge the solution space of the MMAD model in order to be able to reduce utilization losses.

We highlight the importance of level loading the entire production structure in Sequence Design for Mixed Model Assembly (SD for MMA). In Chap. 3, we outline the advantages of using variable takt in combination with sequence planning, or *sequencing*.

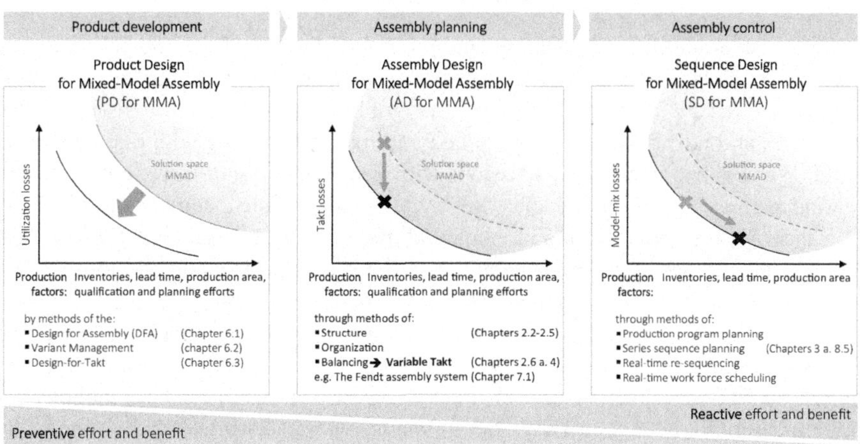

Fig. 9.1 The mixed-model assembly design (MMAD)

In doing so, we propose a combination of the supply-oriented approach using sequence rules and a workload-oriented approach. In this way, supply chains and the load on the workforce can be levelled. A workload-oriented series-sequencing (Sect. 3.2.3) is the current trend at all automotive manufacturers. If variable takt is also used in this phase (in sequencing), even greater variance can be managed and productivity gained. We describe approaches to this in Chap. 3 and in particular in Sect. 8.6.4.

In this book, we specifically focus on the *planning of assemblies*, on Assembly Design for Mixed-Model Assembly (AD for MMA). With Chaps. 2 and 4, the concepts of structure, design, organization, and balancing form the content focus of our book. A highlight is the analysis of the Fendt assembly system in Sect. 7.1, where different concepts from AD for MMA are used. Thus, an extreme amount of variance in and between products is mastered in a successfully operating assembly (Fig. 9.1).

An important contribution is made by our considerations and concepts on takt time groups in Chap. 5. With the developed Variable Takt Time Group Algorithm (VTGA), we present for the first time a quantitative method for the classification of a product or order spectrum into different takt time groups. With the developed qualitative VTGTP (*V*ariable *T*akt time *G*roup *T*ransformation *P*rocess) from Sect. 5.4, we provide an efficient method to examine the product portfolio of each assembly for the ability to form takt time groups and to use variable takt. The VTGTP can be used to determine the key parameters for implementing variable takt in an assembly. In the following case study in Sect. 5.5 on Canyon Bicycles, we successfully apply the VTGTP outside of an automotive assembly and can envision productivity gains of 30% or more.

9.3 Feedback from Business and Science

We have been working, researching, and promoting the use of variable takt for over 5 years. In the course of this time, we have exchanged ideas with numerous people responsible for a wide range of different assemblies. For example, we discussed the properties and advantages of variable takt and matrix assembly with planners from Audi. We also had intensive discussions with a manufacturer of large industrial and marine engines. We discussed and compared the takt module assembly (Sect. 8.3) in Porsche's Taycan production and integrated the concept of variable takt into it. We planned and simulated the implementation of the VarioTakt in a non-automotive, but high-volume assembly at Canyon Bicycles. Not to be counted are the numerous discussions and benchmark visits to our Fendt tractor production facility. One thing that was common to all discussions, after initial skepticism and several (cognitive) understanding loops, all discussion partners were surprised by the simplicity of implementing variable takt. Even if "old hands" who had worked with a fixed takt for decades were often only convinced of the functionality and robustness of the variable takt with simulations.

The successes of Fendt and Fendt's production have been confirmed in recent years with various awards. The World Economic Forum recently recognized Fendt as the "Lighthouse 4th Industrial Revolution" in 2020. In 2017, at the beginning of the implementation of the variable takt, the Industrial Excellence Award confirmed Fendt's future-oriented production management with a Laureate award. And last but not least, the awards from the market: Fendt increased its production volume of tractors by over 50% during these years.

In Sect. 1.1.3, we listed our main research questions, in the course of this book all could be answered. We worked out the advantages of continuous flow production and the limits of the fixed takt in Chap. 2. The benefits and functioning of levelled build-to-order assembly are followed in Chap. 3. We focused in Chaps. 4 and 5 on the functioning, prerequisites, advantages, and experiences with variable takt and the VarioTakt. We showed the advantages of the Fendt assembly system in Sect. 7.1, the Canyon Bicycles case study in Sect. 5.5, and the simulation using automated balancing in Sect. 8.6. We outlined how product design can support flow assembly in VarioTakt in Chap. 6. In the last two chapters, Chaps. 7 and 8, we were able to compare and combine current concepts for managing variance in variable takt assemblies.

A widespread implementation of variable takt is still far away, so the experiences and successes reported in this work can only represent the start of a long (research) journey. With a few exceptions, we presented these publications at the relevant points in this book, no in-depth scientific work has yet focused on the application of variable takt as we describe it here. A broader and, above all, more diversified reappraisal should follow. For example, the VTGA (Variable Takt time Group Algorithm) could be expanded with a stronger practical focus. With the scientific monitoring of further implementations of the VTGTP (*V*ariable *T*akt time *G*roup *T*ransformation *P*rocess), this could be expanded and improved. The boundary between the necessary costs of a re-balancing and potential efficiency gains should become quantifiable, methods of a break-even analysis would be very helpful.

Future research should not only focus on costs but also on the employee, the worker in the assembly line. Further concepts need to be developed to increase the productivity of assembly lines with a wide range of individualized products and, at the same time, to reduce and level out the physical and mental strain on employees. In the end, an assembly line will certainly require one thing above all else for more than a decade: healthy, capable, and motivated employees who work in and for assembly lines.

9.4 Extension of the Toyota Production System with Variable Takt

In this summary, we take the liberty of expanding the house of the Toyota Production System to include the element of variable takt and its most important goals. We do not presume to question the concepts of the Toyota Production System (TPS). Too often its superior effectiveness has been demonstrated in many industries. With our findings from the first three chapters, we make an attempt at the expansion in the context of current and future demands on manufacturing companies and add the elements of variable takt to the TPS house.

We are expanding the "magic triangle" of quality, costs and delivery service to include a fourth target element: *customized products* (Fig. 9.2). This is because, as already described and justified several times in the previous chapters, customers will

Fig. 9.2 Extension of the visualization of the Toyota Production System to include the elements of variable takt for mass customization (adapted from Liker, 2021)

in the future increasingly demand products tailored to their individual needs, in the best quality, at the lowest cost, and in the shortest delivery lead time—mass customization (Sect. 1.2.1; Koren, 2021).

A *harmonized*, and thus *uniformly levelled workload* despite individual products in a build-to-order sequence in production, will become significantly more important as a third employee-oriented target element.

Variable takt significantly expands the solution space in the just-in-time system. With it and other elements from the AD for MMA, a levelling, *heijunka*, of the upstream *supply chains* and the *workload* in production can take place.

Also exciting in this context would be the question of the generalizability of variable takt: for other industries or services. In the meantime, the ideas and principles of TPS have already been successfully transferred to administrative processes. Could the principle of variable takt also contribute to a levelling of the workload in these processes?

9.5 Get Started!

Fendt is unique! More precisely, Fendt's tractor production, because no other assembly plant produces eight extremely different products in terms of cubature, technology, and assembly effort in a continuously flowing assembly line in quantities exceeding 20,000 units (Sect. 4.1). Variable takt has been known in the scientific literature for over 40 years (Kilbridge & Wester, 1962), but there is a reason why the first practical and large-scale implementation developed in Fendt's tractor production: It was necessary! In Sect. 4.3.1, we report on the background to the emergence of the VarioTakt when the Fendt 1000 Vario was introduced and why there was actually already a use for variable takt in the apron conveyer (Sect. 4. 6.1)—which could be adapted at any time by other assemblies that only have an apron conveyer or chain conveyer available.

Not only variable takt can be used to master variance in assemblies, other concepts, above all the currently popular matrix assembly, have been discussed in science and practice for several years. In Chaps. 7 and 8, we present some of these concepts and compare them with variable takt. But not only a comparison, much more interesting seems the smart combination of these different concepts from AD for MMA. The already mentioned Sect. 7.1 with the Fendt assembly system can be seen as a blueprint for a transfer to other assemblies. We were able to show that the application range of a matrix assembly is significantly reduced by using variable takt in an assembly line (Sects. 7.2 and 8.1)—the advantages of a flow assembly remain! The break-even point for conversion to matrix assembly is again shifted in favor of line assembly. Not to mention the significantly lower complexity in planning and controlling a line assembly.

Hybrid assembly can be excellently combined with variable takt and the VarioTakt, as we presented in Sect. 8.2. Especially assemblies with large local takt time spreads can thus specifically master local variance of individual options, e.g.,

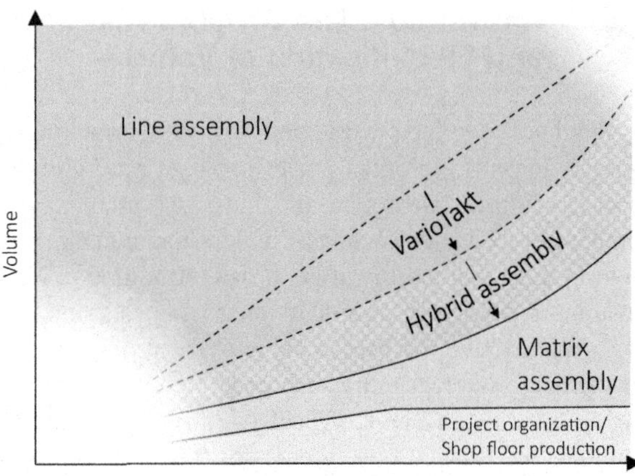

Fig. 9.3 Expansion of the area of use of a line assembly through variable takt and hybrid assembly

through electric drive concepts. The economic application range of a line assembly can thus be significantly increased (Fig. 9.3).

We recommend our experiences with the introduction of variable takt in Sect. 4.8 to all those willing to implement it. The most important hurdle remains to convince all those responsible, the employees involved, and the works councils of the advantages and the mode of operation of variable takt. Much of the content of this work can be used for this purpose, copying (and understanding) encouraged! As the responsible manager, you could specifically create situations in which variable takt plays out its advantages: Add additional products to your assembly line, merge two different assembly lines, significantly increase your productivity goals, or convince the affected employees of a significantly levelled workload, no matter what product-mix the sales department wants.

Our most important hypothesis is why the variable takt is not yet dominant in all assemblies today: It is not necessary! Or rather: It is not ***yet*** necessary. But soon it will be in more and more assemblies. The increasing individualization of products and the resilient distribution of production across several locations are pushing the classic fixed takt solutions to their economic and practical limits (Sects. 1.2 and 1.3). Assembling companies of customized products will have to deal with variable takt, we think that there will be no way around it. With the increasing expansion of flexible transport systems (even if they are not absolutely necessary), the increasing necessity and (perhaps also) with the help of this book, variable takt will find its successful spread.

9.6 An Outlook: The Variable Takt in the Age of Digitalization and Electrification of Vehicles

In this book, we have explained how the variable takt can help level out takt time spreads for entire product portfolios, e.g., products with e-drives. From a necessity comes a virtue. However, the introduction of e-vehicles poses quite enormous challenges to dealer networks, as less maintenance and spare parts are in demand, putting their profitability at great risk (Automotive News, 2021a). At the same time, customers favor the option to order and customize their vehicles online. In the U.S. automotive industry, today's trend is for manufacturers to build digital platforms to interact more with customers and to offer additional services—including Software-as-a-Service (SaaS)—throughout the product lifecycle. (Retailers do not have the financial resources, nor the technical means, to do this. New incentive models to compensate for lower revenue are already in use by BMW and Mercedes-Benz in the US.) Digital business transformation is moving away from the marketing push of product innovation in dealer showrooms to customer pull and the systematic building of lifelong customer relationships (Fader, 2021). We see something similar in other industries, i.e., pharmaceuticals. The variable takt allows companies in this environment to think and act from the customer's point of view in an unprecedented way. We, therefore, take the liberty of once again adapting our already modified TPS house from Fig. 9.2 to this and adapt the "magic target triangle" of quality, costs, and delivery service to mass individualization in Fig. 9.4 (Sect. 1.2.1; Koren, 2021).

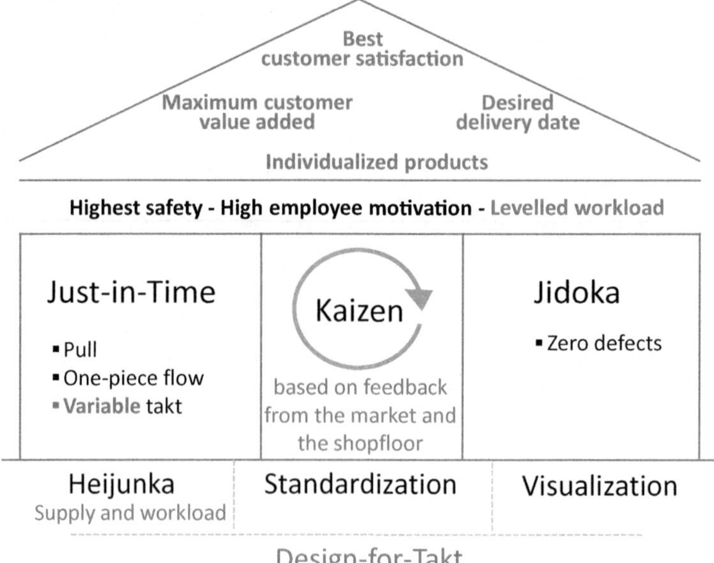

Fig. 9.4 Modification of the visualization of the Toyota Production System to include the elements of variable takt for mass individualization (adapted from Liker, 2021)

From Best Quality to Best Customer Satisfaction Products are manufactured individually without restrictions and only according to the explicit wishes of the customer; this significantly increases customer satisfaction. This new orientation is to be regarded as a higher corporate goal or business unit strategy than the "restricted" view of product quality. Quality is a "must," only through the highest customer satisfaction does sales or company growth occur (through the sale of products *and* services), this is now in the foreground and is considered an immediate and market-based feedback to the entire organization.

From Shortest Delivery Lead Time to Desired Delivery Date The direct contact between customer and manufacturer allows either an immediate dispatching of orders according to the first-in-first-out method or a targeted dispatching on the exact desired date. Equal distribution of orders for reasons of workload levelling is no longer necessary with the variable takt. No other production system is able to show such short and customer-oriented delivery lead times.

From Lowest Cost to Maximum Customer Added Value We have described how, on the one hand, investment costs are minimized and, on the other hand, only real orders from customers are produced. Since the levelling of the line happens automatically and no product-mix rules have to be followed, no phantom orders are produced (with possibly low or negative margins or increased inventory costs), nor is demand aggregated over a planning period. On the contrary, exactly *the right* product with *the right* option variants can be individually produced immediately. Companies can "gold-plate" this dominant Time-to-Market and Customer Centricity. Added value is decisive, not exclusively low costs.

Individualized Products Since each order is given exactly the time it takes to assemble it, even complex orders can be manufactured in any sequence. Thus, more freedom can be granted in the selection, combination or design of product options. Individualized production by product-mix, volume, and options becomes possible. Toyota announced that in the future it will develop its vehicles around an operating system, which dynamically adjusts driving comfort, in order to (passively) individualize its products (Automotive News, 2021b).

All in all, the variable takt allows companies to be more customer-focused and to adapt more quickly to changing market or supplier conditions. Continuous improvement, Kaizen, should be based on market and shopfloor feedback to develop processes in a customer-centric way. It is no longer solely about transactions in a mass market, but about the continuous creation of value for the customer. This leads to increased resilience, performance and at the same time promotes more sustainable growth.

9.7 An Outlook: Where Is the Journey Going for Assemblies

... mainly in one direction: increasingly individualized and structurally different products will have to be assembled on an assembly line. Although software will increasingly take over the task of individualization, the variance between and within products will increase significantly. This variance must not be avoided, it must be mastered—and economically!

Global variance, due to different products in an assembly line, and local variance, due to more diversification in products, will increase for all assemblies. Assemblies will become even more decentralized to minimize global risks, but also in response to customer demand for sustainable and local production. This will lead to smaller, standardized assemblies that will have to economically map a larger number of different products. Especially the established industrialized countries, with their high labor costs, need new solutions to maintain—and continuously increase—productivity despite increasing variance. The approaches of Lean Management are and will remain the basis of successful production. They are the basis of every effective and efficient production concept, only there should or can be less focus on the pure standardization of products. In the end, it is always about balance—variable takt creates this without having to intervene in the design of the products. Thus, the development of further and improved methods to economically master this variance is needed. Variable takt and the VarioTakt should be just one of the possible approaches.

For Fendt, the pioneer of variable takt and name giver of the VarioTakt, the past years of introducing and implementing variable takt have been very successful. The volume was increased, the quality improved and the variance of the tractors offered and their options expanded. New tractor models, expanded digital value chains and services, and the merging of on-board and off-board applications will be cornerstones of future success. Tractor assembly will have to expand its assembly system due to the continuously expanding variance in options and functions, but also due to the upcoming electrification of tractors. The VarioTakt remains one of the most important elements in the Fendt assembly system but will be expanded to include the potential of digitalization and hybrid assembly structures (Sect. 8.2).

Assembly is the place of value creation and customer individualization, it is manual work (at least for a long time yet) and it is for many people the center of their professional life. There is value in it, working with it or for assemblies—to equip you for your future tasks through good management, hands-on research, or very small, daily improvements. With this initial book about variable takt and VarioTakt we explain how it works, convince with many advantages, offer approaches of further scientific work, and above all generate motivation for implementation. It is easier than you think, ***get started!***

References

Automotive News. (2021a). *U.S. dealers are exceptional, but they'er facing big threats*. Accessed July 07, 2021, from https://autonews.com

Automotive News. (2021b). *Toyota eyes personalized updates for better driving*. Accessed July 14, 2021, from https://autonews.com

Fader, P. (2021). *Customer centricity* (150 p.). Wharton School Press.

Kilbridge, M., & Wester, L. (1962). A review of analytical system of line balancing. *Operations Research, 10*(5), 591–742. https://doi.org/10.1287/opre.10.5.626

Koren, Y. (2021). The local factory of the future for producing individualized products. *The BRIDGE, 51*(1), 100 p. https://www.nae.edu/251191/The-Local-Factory-of-the-Future-for-Producing-Individualized-Products

Liker, J. K. (2021). *The Toyota way* (2nd ed., p. 449). McGraw Hill.

Indices, Parameters and Decision Variables

a_j, a_r, a_q	Assembly content of orders *j, r, q* at each station
$a_{j,w}$	Assembly content of order *j* for worker *w*
A_j	Total assembly content of order *j*
α	Lower limit in workload equilibrium as a function of β; relative value
β	Maximum allowed overtakting; relative value
C_b	Actual capacity required
C_v	Capacity provided
D_i	Distance to the next unit of an order from takt time group *i* or launching distance of the orders from takt time group *i*
D_j	Distance to the next unit of the order *j* or launching distance of job *j*
E_{ma}	Design-for-Assembly index
E_{AL}	Balancing efficiency of the assembly line
E_{WSn}	Balancing efficiency of workstation *n*
e_a	Total allowance; time that must be mapped in the assembly line
e_{fa}	Factual allowance; time that must be mapped in the assembly line
e_{pa}	Personal allowance; time that must be mapped in the assembly line
$h(A)$	Relative frequency of operation time of job *A*
Υ	Maximum drift range
i	Index of takt time groups
I	Total number of takt time groups
j	Index of orders
J	Total number of orders
l	Index of assembly operation per station
L_n	Index set of assembly operations *l* assigned to station *n*
m	Index product group, one or more products can be combined into one product group
M	Total number of product groups
n	Index of workstations of the assembly line
N	Total number of workstations in an assembly line
$N_{\min}$	Theoretically lowest number of parts (see DFA)
ot_l	Operation time of assembly process *l*
θ_i	Overtakting limit of takt time group *i*; relative value

(continued)

P. Bebersdorf, A. Huchzermeier, *Variable Takt Principle*, Management for Professionals, https://doi.org/10.1007/978-3-030-87170-3

OV1, OV2, OV3	Short form for option variant 1, 2 and 3 of a product
P1, P2, P3	Short form for product 1, 2 and 3
P1a, P1b, P1c	Short form for product 1 with option *a*, *b* or *c*
P_l	Probability of occurrence of assembly operation *l*
p_i	Proportion of the number of orders in takt time group *i* in relation to the total number of orders *I*
q	Index of orders
Q	Index of the largest order not yet assigned in the VTGA
r	Index of orders
R_n	Weighted workload at station *n*
$R_{w,i}$	Weighted workload of worker *w* in takt time group *i*
s	Station index
S	Total number of stations in an assembly line
TA	Takt time gap
t_a	Standard assembly time of a part (see DFA)
t_{ma}	Determined time for the total assembly of a part (see DFA)
T	Takt time/cycle time
T_i	Takt time of the takt time group *i*
T_j	Takt time of order *j*
T_{Vario}	Takt time within VarioTakt
V_{CS}	Flow velocity of the assembly line, the conveyor system respectively
V_n	Flow velocity at station *n*
w	Index of workers on an assembly line
W	Total number of workers on an assembly line
wd	Workforce density
wlo	Workload per unit length on the assembly object (*w*orkload unit *l*ength *o*bject)
wls	Workload per unit length of assembly access surface (*w*orkload unit *l*ength assembly access *s*urface)
X_{ji}	Binary variable of order *j* (not) assigned to takt time group *i*

Abbreviations

AD for MMA	Assembly Design for Mixed-Model Assembly
AGV	Automated guided vehicle
AR	Augmented reality
ARC	Assembly Revolution Cell
BOM	Bill of materials
CIP	Continuous improvement process
CPS	Cyber physical systems
DFA	Design-for-Assembly
k.o.	Knockout
FLDP	Flexible layout design problem
IAA	International Motor Show in Germany (Internationale Automobilausstellung)
IT	Information technology
IOT	Internet-of-Things
I4.0	Industry 4.0 (The fourth industrial revolution)
JIT	Just-in-time
JIS	Just-in-sequence
AI	Artificial intelligence
CIP	Continuous improvement process
LM-structure	Line-matrix structure (variant of a hybrid assembly)
LML-structure	Line-matrix-line structure (variant of a hybrid assembly)
MES	Manufacturing Execution System
ML-structure	Matrix-line structure (variant of a hybrid assembly)
MMAD	Mixed-Model Assembly Design
MMAL	Mixed-model assembly line
MTM	Methods-Time Measurement; is a method for analyzing workflows and determining target times
MTB	Mountain bike (Canyon Bicycles product group)
NP-hard	Non-polynomial, i.e. exponential computation time is required
PDCA	Plan-Do-Check-Act cycle, also known as Deming cycle

(continued)

P. Bebersdorf, A. Huchzermeier, *Variable Takt Principle*, Management for Professionals, https://doi.org/10.1007/978-3-030-87170-3

PD for MMA	Product Design for Mixed-Model Assembly
PLC	Programmable logic controller
POC	Proof of concept
REFA	Reich committee for working time determination
RWTH	Rheinisch-Westphalian Technical University (Aachen, Germany)
SaaS	Software as a Service
SWS	Standard worksheet
SD for MMA	Sequence Design for Mixed-Model Assembly
SMED	Single minute exchange of dies (method for reducing setups/changeover)
TPS	Toyota Production System
TU	Time unit
TUM	Munich University of Technology (Munich, Germany)
WATT	Weighted average takt time
WHU	WHU - Otto Beisheim School of Management (Vallendar near Koblenz, Germany)
WO	Work operation or activity module

Printed in Great Britain
by Amazon